CW00793718

The New Naturalist

A SURVEY OF BRITISH NATURAL HISTORY

THE SOIL

The aim of this series is to interest the general reader in the wildlife of Britain by recapturing the enquiring spirit of the old naturalists. The editors believe that the natural pride of the British public in the native flora and fauna, to which must be added concern for their conservation, is best fostered by maintaining a high standard of accuracy combined with clarity of exposition in presenting the results of modern scientific research.

The New Naturalist

THE SOIL

B. N. K. Davis, N. Walker,
D. F. Ball and A. H. Fitter

With 14 colour photographs and 100 black and white photographs and drawings

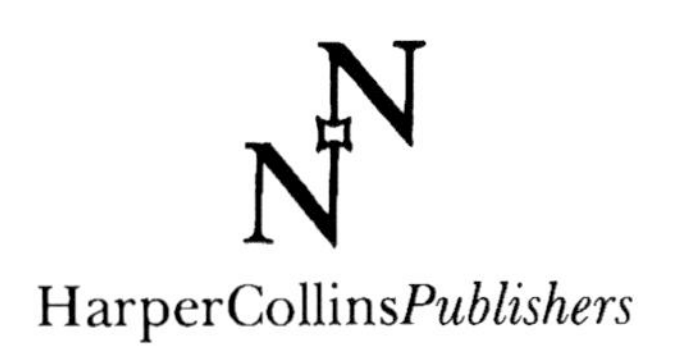

HarperCollins*Publishers*

HarperCollins*Publishers*
London · Glasgow · Sydney · Auckland
Toronto · Johannesburg

First published 1992

ISBN 0 00219903 3 (hardback edition)
ISBN 0 00219 904 1 (limpback edition)

Printed and bound in Great Britain by
Butler & Tanner, England

This edition is produced from an original copy by William Collins.
For further information go to www.newnaturalists.com

Contents

Editors' Preface

We are all familiar with the plants and animals with which we share the world above the soil surface. We are much less familiar with the inhabitants and processes in the ubiquitous but inaccessible world of the soil. The organisms in this hidden world are often microscopic, small enough to live in the maze of narrow pores, and are concealed by the opaque matrix in which they live. Taxonomic difficulties add a further barrier to exploration. *The Soil* is an account of the soil as a living system, in which moles, minerals, molecules and microbes interact with vegetation, under the influence of climate and man.

The soil, the earth's skin, has been scarred, peeled off or incurably bruised over much of Britain. It is all too easy for heavy machinery to destroy in minutes a soil profile that has taken centuries to develop. Perhaps this book will help us to appreciate those precious sites where its structure remains intact. It also shows how, if compaction can be avoided, heavy machinery can have a positive role in the large-scale restoration of soil capable of supporting vegetation on mine or industrial waste.

Processes in the soil matter to farmers and gardeners, moles and oak trees. To understand them is a necessity for agronomists, an aspiration of ecologists, and a pleasure for others. Many policy decisions on agriculture and conservation depend on the ability to understand and manipulate soil processes. This is an area where informed public opinion has an important part to play in moulding public policy. Those who read this book will be in a better position to understand some of the more controversial aspects of man's impact on the soil.

Research has advanced on many fronts since Sir John Russell wrote *The World of the Soil* in this series in 1957. Until now many of those advances have been accessible only to specialists. The authors of this book are in a unique position to introduce us to this new material and to provide an exceptional guide to the ecosystem we tread on every day. It is a credit to the subject that it has advanced far enough in 34 years to merit this new approach, and a credit to the series that it has lived long enough to include this second book.

Authors' Preface

Soil science came of age in Britain with the publication in 1912 of the first edition of E. J. Russell's *Soil Conditions and Plant Growth*. Forty-five years later, after retiring as director of Rothamsted Experimental Station, Sir John Russell published *The World of the Soil* in the New Naturalist series to bring the burgeoning subject to a wider audience. "Fifty years ago" he wrote in his preface, "this book would have been much easier to write than it has been today. Enough was then known about the wonders of the soil to show that deeper mysteries lay beyond. The facts gleaned were simple, the generalisations were broad and easily comprehended...Now it is all very different. Vast numbers of learned memoirs have been written about [the subject]." If this statement were true then, how much more so today. The subject has continued to grow in breadth as well as depth with several international journals on soil biology starting since 1957 and a string of published symposia. *Soil Conditions and Plant Growth*, revised by E.W.Russell and now edited by A. Wild, is in its 11th edition (1988), and is still a standard reference work.

With growth has come increasing specialization – in soil physics, pedology, soil biology, microbiology and applied aspects, such as agriculture and land restoration. The student and professional scientist is now well supplied with monographs on all these topics but there is increasing need to see the world of the soil as a whole; to appreciate the complex and delicate structure of this thin skin on the earth's surface, the activities of microorganisms in the cycles of decay and renewal, the interwoven lives of animals and plants below and above the soil surface, and man's ability to use or abuse this vital resource. It is difficult to do justice to this broad spectrum, and we have had to be selective in our coverage. We have tried to illuminate a wide range of themes which interest us and which we thought would interest the non-specialist reader. Inevitably, this selection leaves many gaps but if we succeed in whetting an appetite for more information we will be well satisfied.

We have tried to be up to date in concepts and discoveries but, even while preparing the book, the agricultural scene has changed radically. To increase productivity has ceased to be the driving force behind so much of the research on soils that has marked the past 50 years. Crops, however, are just one expression of a soil's potential: to create and manage diverse ecosystems is no less a challenge; amongst other things, it entails research on reducing soil fertility.

Each section of the book has been read and criticized among ourselves but, in addition, comments have been obtained from outside experts on particular topics: micro-arthropods – M. Luxton, woodlice – P. T. Harding, millipedes and centipedes – R. E. Jones, spiders – E. Duffey, ants – T. J. King, other insects – R. C. Welch, earthworms – B. M. Gerard, nematodes – K. Evans, snails and slugs – B. Eversham, agriculture – B. G. Davies, land restoration – S. G. McRae. We are most grateful to all these, and hope to have avoided errors even if we have not presented a topic in the way they would have done.

The sources of illustrations are given with the captions and we would like to thank the following for providing photographs: J.M. Anderson, T. Bauer, Broom's Barn Experimental Station, A.F. Brown, G.P. Buckley, J. Day, B. Dickerson, K. Evans, R. Evans, R.D. Finlay, Frank Lane Picture Agency Ltd, J.A. Grant, Christine Hepper, Dick Jones, T.P. McGonigle, S.G. McRae, R.H. Marrs, J. Miles, National Museum of Wales, J.A. Thomas and T.C.E. Wells. S.V. Green and R.C. Welch kindly made original drawings of snails, woodlice and beetles. Figures 45 and 46 were drawn by Paul Joyce. C.A. Howes generously provided original data on mole distributions.

We gratefully acknowledge the kind permission of the following journals and institutes to reproduce copyright illustrations: *Pedobiologia* (Figs 1, 31), *Journal of Soil Science* (Fig. 6), Blackwell Scientific Publications Ltd (Fig. 7), Plenum Publishing Corporation (Fig. 9), Soil Survey and Land Research Centre (incorporating the Soil Survey of England and Wales) (Fig. 11), *Transactions of the British Mycological Society* (Fig. 16), *Naturwissenschaften* (Fig. 23), *Behavioural Ecology and Sociology* (Fig. 30), *Netherlands Journal of Zoology* (Fig. 39), John Wiley and Sons (Fig. 45), *Geoderma* (Fig. 49), North Holland Publishing Company (Fig. 50), Institute of Terrestrial Ecology (Fig. 51), *Journal of Ecology* (Fig. 52), Grassland Research Institute (Fig. 57), Central Electricity Generating Board (Fig. 67), Ready Mixed Concrete (United Kingdom) Ltd (Fig. 69).

Finally, I should like to thank K. Mellanby for his encouragement in writing this book (B.N.K.D.).

Units of measurement

Science now uses metric units almost exclusively but it is often easier for many people to visualize an acre rather than a hectare, or to appreciate altitude in feet rather than metres. We have therefore given both in many instances, especially when dealing with historical data. In some cases, English units have become disused except by a few but it would be inappropriate to quote wheat yields in the Middle Ages in anything except hundredweights.

1,000 microns = 1 millimetre (mm)
25.4 millimetres = 1 inch
1 metre (m) = 39.37 inches = 3.28 feet
1 hectare (ha) = 2.47 acres
1 kilogramme (kg) = 2.20 pounds
1 tonne (t) = 1000 kilogrammes = 2205 pounds
1 hundredweight (cwt) = 112 pounds = 50.8 kilogrammes

1

Architecture of the Soil World

We live on the rooftops of a hidden world
Peter Farb 1959

It is difficult to visualize the world of the soil as it appears to a worm or a woodlouse, a mole or a microbe. We may dig a pit in a woodland, grassland or arable soil and describe the different sections exposed to view; or feel the distinctive textures of a peaty moorland soil and a sandy heathland soil. We can measure the sand, silt and clay contents, or analyse a soil for its available plant nutrients – nitrogen, phosphorus and potassium. From our perspective, such apparent abstractions are a necessary step towards understanding a soil, but this is a long way from knowing how an individual plant rootlet will behave as it grows; or how a parasitic eelworm makes its way through the soil to attack the growing root.

Partly, it is the three dimensionality of the soil environment, and partly its physical complexity and scale, which are beyond our direct experience. Quite apart from gravel and larger pieces of rock, there is more than a thousand-fold range in size between the two extremes of the spectrum of what soil scientists call 'fine earth': between coarse sand particles, up to 2mm in size, and those of clay minerals which are less than 0.002mm. (The sizes of the soil inhabitants cover an even larger spread, between a 200mm earthworm and a 0.002mm microbe, for example). These mineral particles, together with an intimate mixture of living and dead plant material, form a spongy matrix permeated by pores filled with air and water. The pores themselves may comprise 30-50% of the total volume in a good topsoil – plenty of space for air-breathing creatures of all sizes, and for those dependent upon an essentially aquatic way of life.

The soil surface and superficial plant remains

The soil environment can be looked at from many viewpoints: as a sequence of approximately horizontal zones of distinctive character and properties from the soil surface downwards; as a medium that provides varying levels of physical and nutritional support to plants with differing requirements; as a fabric affording many and varied niches suitable for particular soil organisms; as a reactive skin covering much of the earth's surface that provides a sink and a buffer for rainfall and for airborne chemicals; and, overall, as a vital resource that sustains life on earth.

These approaches are returned to in future chapters, but in considering the architecture of the soil world it is convenient to start on its roof – at the soil surface – for this is the part that is easiest to observe, and hence is most familiar. A variety of creatures are found simply by turning over stones and

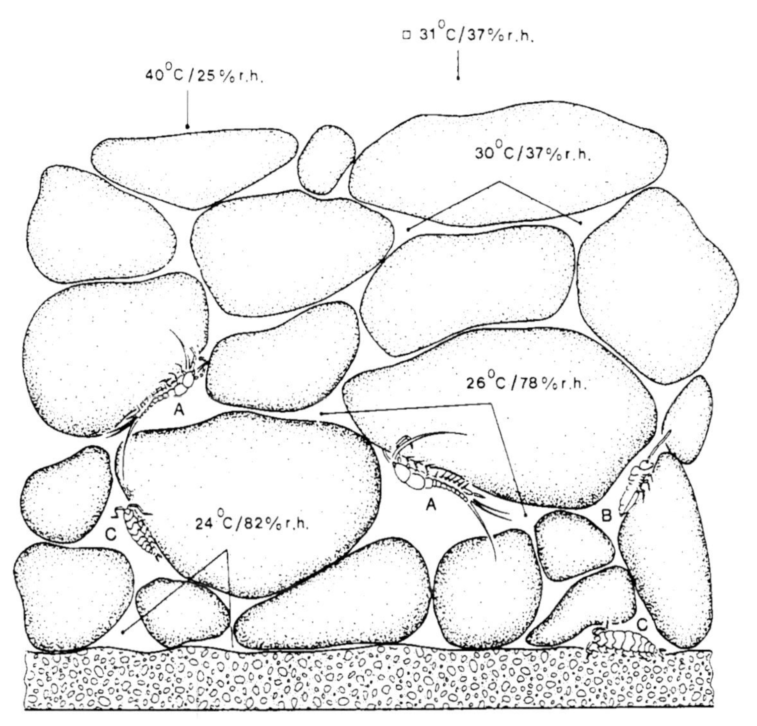

Fig. 1 Microclimate: temperature and humidity gradients in a stony habitat in strong sunlight, and the effects on positions occupied by various small arthropods; A = bristle-tail, B = springtail, C = woodlouse. The temperature decreases from the top downwards while the relative humidity increases. When the stones are shaded, the bristle-tails move to the underside of the upper stone layer. (Adapted from G. Eisenbeis 1983.)

logs, and while some of these may not strictly be called soil animals, yet they are often very dependent upon the nature of the underlying material at some stage of their life. Many are nocturnal, and merely shelter here during the day to avoid desiccation or predation by birds. These include predatory ground beetles and spiders, and vegetarians such as woodlice and slugs. Stones that are smaller than about 10cm are of little value for providing special microclimates unless they are scattered quite thickly over the ground as in shingle, in a quarry or on a mountainside; Figure 1 shows the temperature and moisture regimes at different depths within a heap of stones inhabited by various arthropods. Large boulders, on the other hand, are usually well bedded into the ground, and if turned over will reveal true subterranean animals such as worms, or the underground galleries of ants.

Cracks and fissures in the ground serve much the same function as stones in affording protection from dry conditions for species that cannot burrow for themselves. Cracks are a common feature of clay soils during the summer, and may penetrate a foot or more in grassland or arable fields.

Plant cover affects conditions on the soil surface in more ways than inorganic objects do because of its more varied and complex structures and

Fig. 2 Leaf litter, twigs and branches in a mixed oak/ash woodland in spring. Note that oak leaves remain but all the ash leaves have disappeared. (Photograph B.N.K.D.)

thermal properties. Adhering vegetation, such as moss and algae, liverworts and lichens, harbours a rich micro-fauna of protozoa, nematodes, tardigrades, small mites and springtails. Larger moss cushions, grass tufts and rosette plants shelter a wealth of small beetles and other insects, both adults and larvae. Dead plant material not only gives shelter but offers food for a wide range of animals and fungi which fulfil a vital role in returning the store of organic material and nutrients back to the soil.

A forest floor has the greatest variety of plant litter – annual sheddings from the tree canopy with occasional branches and logs, all in varying stages of disintegration and decomposition (Fig. 2). The total litter fall in a deciduous woodland in this country is around 2.5 kg a square metre (10 tons an acre). Seventy-five percent of this is leaf litter, the rest being made up of twigs, bark and seeds etc. The detailed character of this litter is important in providing habitats and food sources for different organisms. In a particular study of woodland litter, H. Heatwole decided there were three main categories. His first class consisted of leaves that roll or bend when they fall to the ground, thus producing large, round or angular spaces between them. Class 2 consisted of leaves that remain flat and so have small, narrow interspaces, and class 3 consisted of solid, woody objects. Each class was subdivided, for example to distinguish thick leathery leaves from thin papery ones and conifer needles, or accumulations of twigs from large logs. Some of these types are characteristic of natural, mixed deciduous woodland, beech hangers or pine woods in this country. A few are perhaps associated more with parks and gardens, where exotic trees and shrubs like rhododendron have been planted.

There is an enormous difference in the persistence and smothering effect of different kinds of leaves, ranging from ash and apple, which disappear in weeks, to beech, holly, rhododendron and conifer needles which may last for years. This difference is partly due to their size and thickness, and partly to their palatability to soil animals and susceptibility to fungal attack. The depth of litter can vary greatly depending on the density of trees, the time of year, the properties of the underlying soil, and the micro-relief of the ground. Mounds and convex surfaces may remain largely bare of litter while leaves and twigs accumulate in hollows which therefore act as foci for litter- seeking invertebrates. The student of these groups soon gets an eye for such 'hot spots'; with experience, he can judge very accurately what species to expect.

Where litter persists for several months or years, one can usually see three distinct organic layers above the mineral soil itself. The uppermost layer of curled and uncompressed leaves has a great deal of interstitial space. This is the zone favoured by large, active springtails which grow to 5-6mm in size, and which form the prey of many beetles and spiders. Both hunting and web-building spiders exploit this open-textured but sheltered environment. The webs may be fairly simple arrangements of criss-cross threads spun across the ends of the rolled leaves, but these suffice to entangle or delay weak prey. Experiments have shown that leaf characteristics influence the numbers of spiders in woodland litter: curled leaf litter tends to support higher densities and a greater assortment of spiders.

Beneath this layer of relatively unaltered plant remains comes a zone of partly decomposed but still clearly recognizable plant fragments, and below this a zone of amorphous, finely divided organic matter. These three layers have been given a variety of names, but it is convenient to refer to them as the L (litter), F (fermentation) and H (humus) layers. The fermentation layer is where most of the litter decomposition takes place. This is the home of several kinds of millipedes, woodlice and fly larvae, some small earthworms, many mites and shorter-springed springtails. These are described in more detail in chapters 4 and 5.

Grassland litter differs from woodland litter in that the dead grass does not fall to the ground in the same way but remains arched over the surface for some months, and only gradually sinks down and disintegrates. Highly siliceous grasses, such as tor grass *Brachypodium pinnatum*, form a distinctive and persistent litter mat which few invertebrates appear able to digest. Even palatable meadow grasses, however, can give rise to a peaty mat on the surface of the ground if earthworms are absent; this was seen in New Zealand, for example, when settlers first converted the native vegetation into pasture land with introduced grasses from Europe.

We can move one step nearer to appreciating the structural diversity of litter if we take vertical sections and view them through a microscope. This has been done in both woodland and grassland by cutting small cores or blocks, impregnating them with gelatine, and slicing them up (Fig. 3 and Plate 1). In a study of woodland soils, J.M.Anderson recognized up to seven main classes of structures in a section through the litter, fermentation and humus layers of a sweet chestnut stand. These included not only various leafy and woody items and the cavities between them, but also plant roots with or

Fig. 3 Section through soil humus impregnated with agar jelly showing the 'primitive' insect *Campodea staphylinus* (Diplura). This species is colourless and blind but has highly developed tactile senses. It does not burrow but moves through the soil cavities using its antennae to locate a pathway when moving forward, and its equally well-developed posterior feelers' (cerci) when moving backward under confined conditions. (Photograph J.M. Anderson.)

without their fungal associations, faecal pellets of various invertebrates, and animal remains. He subdivided several of these classes to produce a total of 24 microhabitat categories which were thought to be significant for soil mites (Table 1). By examining many sections in a standard way, it then became possible to relate the number of kinds of mites identified in a section with an index of diversity computed from the number of microhabitats present. The richest zone tended to be the fermentation layer which attracted about 21 species of oribatid mites, whereas the humus layer below only supported some 10 species.

This approach is clearly an advance over macroscopic analyses of litter habitats but many important criteria are still left out. We could perhaps identify the kinds of leaves present, up to a certain stage of disintegration, but their chemistry and relative palatability – the presence of sugars, cellulose, waxes, tannins and lignin – would still elude us. It is as if we tried to distinguish between caster sugar and salt by eye alone.

Humus

We should think of this surface litter and decomposing organic matter, not just as an inert physical habitat for mites and other organisms, but more like the house which Hansel and Gretel found. This, you will remember, was

Table 1 Microhabitat categories used in soil structure analysis (Adapted from J.M.Anderson, 1978)

1. Intact leaves
2. Leaf fragments ≥5mm
3. Leaf fragments ≥1mm <5mm
4. Leaf fragments <1mm
5. Humus/soil organic matter
6. Twigs
7. Wood fragments ≥5mm
8. Wood fragments ≥1mm <5mm
9. Wood fragments <1mm
10. Roots without fungi (mycorrhiza)
11. Roots with living mycorrhiza
12. Roots with dead mycorrhiza
13. Other macrophyte material
14. Clear fungal threads
15. Brown fungal threads
16. Faeces: free living macrofauna
17. Faeces: plant-inhabiting macrofauna
18. Faeces: free living mesofauna
19. Faeces: plant-inhabiting mesofauna
20. Animal remains
21. Mineral material
22. Cavities ≥5mm
23. Cavities ≥1mm <5mm
24. Cavities <1mm

made of ginger-bread, chocolate and barley sugar. The various components of fresh litter – fruits, leaves, stems and bark – differ greatly in their chemical make-up, and these differences are reflected in their rates of breakdown. The soft parts of leaves, containing sugars, proteins and starch within the cells, are quickly attacked and digested by earthworms, millipedes, springtails and other soil animals. One can often find a perfectly skeletonized leaf in which every vein has been left intact after removal of the lamina by microarthropods. The veins and other more woody structures are largely composed of cellulose. Snails are among the few animals that can secrete cellulase and so digest cellulose directly. Most cannot digest cellulose until it has been chemically shredded by microbial attack into simpler molecules. A few animals, ranging from termites to cattle, have evolved the trick of employing microflora in their gut for this purpose. The toughest woody fibres are composed largely of lignin which is highly resistant even to microbial attack, and these therefore remain intact for a long time. Recalcitrant, too, are waxes and resins, as can be seen in the persistence of holly leaves and pine needles.

The role of soil microorganisms in plant decomposition is described in chapter 6. At a simple level, the relative rates of breakdown of plant structures reflect the ratios of carbon to nitrogen in their chemical make up. Grass leaves have a carbon : nitrogen ratio of about 5:1, barley straw about 60:1 and pine needles about 100:1. Since soil microbes have a low carbon : nitrogen ratio of 7 to 6 or even 4:1, it follows that they cannot fully exploit plant tissues that have higher proportions of carbon without drawing on other sources of nitrogen. This has practical implications in the case of straw that is incorporated into the soil after harvest, a point discussed further in chapter 8. Similar arguments apply to the carbon : phosphorus and carbon : sulphur ratios, though the supplies of phosphorus and sulphur are not so limiting.

Humus is the final product of organic matter decomposition. It is a dark amorphous material consisting of complex organic molecules which can be broken down into humic and fulvic acids. There are, therefore, three interrelated organic fractions in soil. First, there are the plant (and animal) residues which form the main source of available nitrogen, phosphorus and sulphur for new plant (and animal) growth. Secondly, there is the microbial biomass which acts as a temporary store of such nutrients, and thirdly, a persistent humus fraction which is highly resistant to further breakdown but which can release nutrients very slowly. Measurements have been made of the age of humus by radio-carbon dating techniques; that is, by measuring the proportion of the radio-active isotope ^{14}C left in the humus, and calculating the time since it must have been taken up by the living plant as carbon dioxide from the atmosphere. Such measurements give periods of several centuries, components such as humic acid persisting for over a thousand years in some instances. To use a monetary analogy, these three organic fractions – plant residues, microbial biomass and humus – might be represented respectively by goods which are traded for cash, a deposit account, and a long-term insurance policy.

The conversion of such stores of nitrogen, phosphorus and sulphur to available, 'cash-in-hand', nutrients is called mineralization; the mineral forms are the ammonium cation (NH_4^+), and the nitrate (NO_3^-), phosphate (PO_4^{3-}) and sulphate (SO_4^{2-}) anions. This important topic of nutrient cycling is developed in chapter 6. The carbon in humus is oxidized to carbon dioxide and lost back to the atmosphere. Usually a steady state is reached between gains and losses of carbon, but under waterlogged conditions plant remains may accumulate as peat. It is worth remembering that the world's vast deposits of coal, oil and gas represent the preserved surpluses of carbon built up by plant tissues in former ages. On the other hand, if peat is dried out and cultivated, as in the Cambridgeshire fens, the stores of carbon are quickly oxidized away again. We are drawing on our capital here just as we are with fossil fuels.

The existence of distinct and persistent litter and fermentation layers characterizes what are called mor humus soils. Mor formation occurs typically on well drained, very acid soils under conifer woodland and heathland. The main deep-burrowing earthworm species *Lumbricus terrestris* cannot survive in such acid conditions, and in its absence the organic matter is not readily incorporated into the underlying mineral soil. In contrast, in well drained, less acidic or calcareous soils, the feeding activities of these earthworms create an intimate mixture of organic matter and mineral soil known as mull humus. This is the typical form of humus found in deciduous woodlands, lowland grasslands and in derived arable soils.

These terms mull and mor were coined in 1878 by the Danish forester P.E.Müller who first recognized their significance as indicators of soil condition and forestry potential. W. L. Kubiena's classic work in 1949 on *The Soils of Europe* recognized 16 main types of humus of which mull and mor represent two extremes. An intermediate condition, which Müller called 'insect mull' but which is now generally known as moder, is characterized by a well developed H layer with thin L and F layers. This condition is attributed to the feeding activity of large arthropods such as millipedes, woodlice and fly larvae in soils with low earthworm populations; sections through samples of the litter impregnated with agar or resin often show a granular structure produced from the faecal pellets of these creatures. Cause and effect, however, are difficult to disentangle in soil ecosystems. The presence of a particular soil faunal assemblage controlling the breakdown and incorporation of organic matter is dependent on the nature of the overlying vegetation and on the chemical and physical properties of the soil itself. An interesting experiment illustrating the interaction between earthworms and different mixtures of trees is described in chapter 5. Such a positive feed-back system is disrupted when there is a major perturbation such as woodland clearance or liming.

The greatest differences between soils, and hence between associated soil faunal communities, are found over the range of natural and semi-natural plant communities. Cultivation obliterates most of the inherent natural differences in humus type and soil chemistry. Under a regime of arable cropping, all or most of the aerial vegetation may be harvested, so there is little or no 'litter' to be returned to the soil. There is, nevertheless, a substantial root biomass which is returned annually; more than would die naturally under a perennial vegetation cover. Indeed, the roots of a barley crop, together with the stubble and chaff, exceed the organic matter in the straw. This explains why clear differences in soil organic matter have seldom been found when comparing fields where the straw has been burnt, with fields where the straw has been dug in (see chapter 8).

Whilst humus plays a relatively minor role in crop nutrition, it plays a very important role in modifying the texture and structure of arable soils. A high organic matter content in soils makes for easier cultivation and better seed beds. This is because of its effects in promoting the stability of soil crumbs. A rich humus encourages earthworms which in turn promote a spongy, porous structure through the intimate mixing of mineral particles and organic matter in their casts.

The mineral skeleton

Some soil types consist entirely or predominantly of organic matter; for example, shallow organic soils may overlie hard unweathered rocks in high rainfall areas, and deep peats occupy basin sites more widely. For most soils, however, it is the mineral component which is dominant. This develops through processes acting either directly on weathered rocks or on unconsolidated superficial geological deposits such as glacial drifts or alluvium. These parent materials differ greatly in mineralogical composition, and the processes of soil development also vary greatly in response to landform, climate and management history (chapter 2). The resultant diversity of mineral soil types and properties is thus even greater than that found within the litter and humus layers.

The relative proportions of different size particles are a key influence on the physical character of soils. Particular attention is conventionally given to the 'fine earth' fraction of the soil – particles less than 2mm diameter which are divided into sand, silt and clay grades. Although gravel and stones may make up a large part of the total soil, it is these finer fractions which are more important in determining soil texture, water movement and chemistry, and in influencing the conditions for soil organisms.

The coarse to medium sands come at the visually distinct range, from 2 to 0.2mm in diameter, their size and angularities rendering them resistant to movement within the soil. They are derived from the the physical weathering of the rocks of the earth's crust: through the shattering effects of heat and cold, from the grinding of rock upon rock in glaciers and rivers, and by the blast of other sand grains transported by the wind.

Very fine sands and silts, between about 0.06 and 0.002mm, are chemically similar but their small size renders them more mobile. In the words of V.I.Stewart, they "readily flow when saturated, clogging pores between

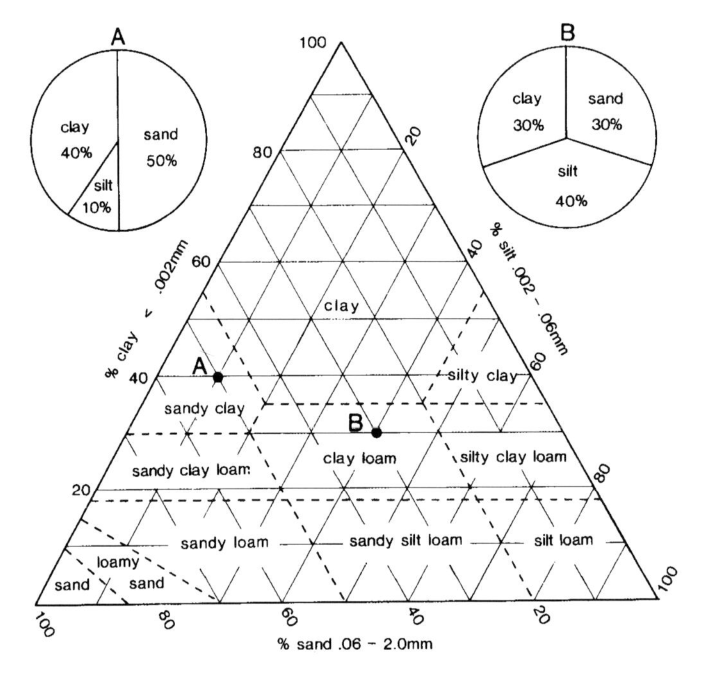

Fig. 4 Identification of soil texture class according to the sand : silt : clay content, and the proportions of each particle size class in a typical sandy clay (A) and a clay loam soil (B). (Based on the classification system for soil textures adopted by the Agricultural Development and Advisory Service in 1984.)

larger particles, in-filling drains and lubricating the erosion of soil downslope... behaving", as he puts it "like a pile of minute billiard balls".

Whereas fine sands are comparable with caster sugar, and silt with icing sugar in particle size, the individual clay particles are an order of magnitude smaller still. Less than two thousandths of a millimetre in size, they are released from sedimentary rocks or derived from primary minerals by chemical weathering. To a large extent, they are the engines of chemical activity in soils, as described below. Physically, clay often dominates the textural characteristics of soils; their downward mobility in undisturbed soils produces distinctive soil types (chapter 2), while their influence on tillage is described in chapter 8.

The size limits of sand, silt and clay particles are thus not purely arbitrary but relate to the observed behaviour of soils and to their feel. The triangular diagram in Figure 4 is the latest of several schemes, used nowadays in both Britain and America, for classifying soil textures according to the proportions of these fractions. Any particular soil can be defined by the percentages of sand, silt and clay that it contains. It is located on this diagram by the intersection of lines parallel with the appropriate sides of the triangle. Another way of viewing the difference between contrasting soil types, such as those labelled A and B in this figure, is to show the relative proportions of sand, silt and clay as circular diagrams. Soils with less than 30–35 per cent clay are called loams – sandy loam, silt loam, clay loam etc – and these

Table 2 Identification of soil textures by hand.

The following key provides a guide to five distinctive soil texture classes. A subdivision of 'sands', 'loams' and 'clays' into the 11 classes in Figure 1.4 can be done in a similar way. Take about a dessertspoonful of moist soil and knead it thoroughly between finger and thumb until the aggregates are broken down. Wet it if necessary until the soil exhibits its maximum cohesion.

1. Is the moist soil predominantly sandy?	Yes →	**2**
	No →	**3**
2. Is it difficult to roll the soil into a ball?	Yes →	**Sands**
	No →	**Light loams**
3. Does soil mould to form an *easily deformed* ball and feel smooth and silky?	Yes →	**Light silt**
	No →	**4**
4. Does the soil mould to form a *strong* ball which smears but does not take a polish when rubbed?	Yes →	**Clay loams**
or Does soil mould like plasticine, take a polish and feel sticky when wetter?	Yes →	**Clays**

include our most common agricultural soils. Some deposits that are particularly rich in clay are quarried for brick making.

Whereas soil analysts can measure the proportions of sand, silt and clay fairly precisely, and then read off the texture from the diagram, the farmer and soil surveyor must often try to reach an opinion in the field by the feel of the soil. A key to the identification of soil textures was revised in 1984 for this purpose by the Agricultural Development and Advisory Service (ADAS) of the Ministry of Agriculture. The procedure is easy enough for anyone to do, and is summarized here (Table 2). The purchaser of a new home may like to know how easy the soil in his garden will be to work, and how it is likely to respond to rain or drought. It needs practice and experience to become fully proficient, of course, but for most practical purposes it is enough to differentiate soils into five classes – Sands, Light loams, Light silts, Clay loams and Clays.

This classification takes no account of soils that have large amounts of organic matter which can greatly modify the way soils behave, so ADAS has included a classification of humose and peaty soils. Organic-rich soils are prevalent in the Fenland of East Anglia, and frequent in areas of high rainfall, mostly over 2000 ft (600 m) in southern Britain but down to sea level in northwest Scotland.

Soil structure refers to the natural aggregates (structural units or 'peds') into which the primary mineral particles and organic matter are formed. The size and shape of these aggregates largely determine the distribution of living space within the soil. Texture and structure in soil are often confused. As Stewart says "when assessing the quality of cloth by its feel we should try to distinguish between characteristics determined by the nature and thickness of the thread and characteristics determined by the nature and closeness of the weave. Similarly, when feeling soil, we should try to differentiate between characteristics determined by the nature and size of the fundamental particles and those determined by the manner of their arrangement." Soil texture is a primary characteristic while soil structure is a secondary and variable feature reflecting the way the soil has been modified by living agencies, including man.

Farmers and gardeners spend a great deal of effort to produce good soil structure, or tilth, and this is much easier with a loam soil than with a heavy clay. Luckily, only a few soils in Britain have more than 50 per cent clay – though this fact is little consolation for those who have to contend with soils on London and Oxford Clays which do. Tilth in clay soils can be improved by digging in organic matter, while marling – the spreading of calcareous loamy soil – was formerly practised very widely on light soils in East Anglia. Where the fenland peats of Cambridgeshire are being eroded and oxidized away, ploughing now brings up silts from below, and creates new soil textures with which to work.

Soils are seldom structureless (i.e. without structural aggregates) except pure sand soils, such as dunes, or recently exposed, 'unripened', sediments, such as the Dutch polders when they were first drained. Grassland or woodland soils of heavy texture often show block shaped or prism-like structural units with visible cleavages between them that allow rapid movement of air and water. At a closer view, one can usually see that the top soil is aggregated into crumbs a millimetre or two in size. These crumbs retain their identity for a period even after washing in water, and have their own internal geometries. One can picture a soil crumb as a collection of fine rock particles and amorphous organic matter held together by packs of clay platelets and various cementing agents. The size and stability of these crumbs depend a great deal on the microbial contribution as well as the amount of clay in the soil; a single crumb of soil may contain a hundred million bacterial cells producing polysaccharide gums, and five metres of fungal mycelium binding the mineral particles together. When thin sections through soil crumbs are examined through a petrographic microscope, one can see many of these features, and calculate that the capillary pores within them may represent a third to a half of the total volume of the crumb.

Clays and soil chemistry

The clay minerals are important for their powerful influence on the physical properties of soils. One acknowledgment of this is seen in the triangular diagram of soil textures shown earlier: almost any soil containing a clay fraction of more than 40% is simply called 'clay', while a soil needs about 90% of sand fraction for it to be called just 'sand'. It is their physico-chemical attributes, however, that really set the clay minerals in a separate category from sands and silts. Their importance in the cycling of nutrients parallels that of the soil fauna in the cycling of energy.

In most British soils, the clay fraction has been derived at second hand by release of clay minerals from sedimentary rocks and glacial drift in which they are already present. Ultimately, of course, the clay minerals are derived from igneous rocks such as granite through weathering over very long periods of time. These clay minerals are composed of two basic units each consisting of a lattice-like sheet. One unit consists of a silicon-oxygen (SiO_4) layer, and the other consists of an aluminium-hydroxyl ($Al(OH)_6$) layer. Owing to the way the atoms pack together, these are called silica tetrahedra and alumina octahedra. The two kinds of sheets are linked together by sharing some of the oxygens, either as a simple pair of sheets (the 1:1 type) or as an alumina layer sandwiched between two silica layers (the 2:1 type). This layer structure, coupled with their very small size (less than 0.002mm), gives clay particles a very large surface/volume ratio; whereas the total surface area of sand grains in a gram amounts to a few hundredths of a square metre, the total surface area of clay particles may lie between 50-300 square metres per gram.

Some 2:1 clays, such as the montmorillonite-vermiculite group, can take up water molecules between the lattice layers and release them again, and are therefore called 'expanding' minerals. Expansion and shrinkage of clays are well known phenomena which can affect the stability of buildings; however, the non-expanding illites, or mica-clays, are the more common 2:1 clays in British soils. The most prominent 1:1 clay on a world scale, kaolinite, is also a subordinate clay in Britain. In its industrial form it is known as china clay and is quarried in Cornwall (see chapter 10).

Some of the silicon atoms in the clay crystals can be replaced by aluminium, and some of the aluminium atoms by magnesium, producing a complex assemblage of different clay mineral 'species'. As a result of these replacements, the clay particles carry a negative charge and so attract positive (basic) ions such as calcium, potassium and ammonium which are held on the surface of the platelets. Since such cations are not tightly bound in the crystal structure they can be readily 'exchanged' with other cations in the soil solution. Different clay minerals have different capacities for ion replacement; thus the relative cation exchange capacity (c.e.c.) of kaolinite, illite and montmorillonite is 1:4:10.

It is this feature that renders the clay minerals a key component of the soil world. They act as a 'buffer' controlling throughput and release of nutrient cations in the way that soil humus acts as an anion (negative ion) buffer retaining and supplying nitrate, sulphate and, to an extent, phosphate ions. Soil humus has a dual role in that it also has a cation exchange capacity as great or greater than that of montmorillonite; it can provide virtually the entire c.e.c. reservoir in sands, and can thus be the critical nutrient store. However, most soils have more clay than organic matter, and clay is a more permanent feature of soil composition, so c.e.c. normally depends heavily on the clay fraction.

The application of ammonium sulphate fertilizer to the soil as a source of nitrogen provides a practical illustration of clay chemistry. The ammonium ions attach themselves to the clay lattice displacing calcium ions in the process. The calcium may remain in soil solution or it may be washed out of the soil as calcium sulphate. When the ammonium is subsequently converted by nitrifying bacteria to nitrate (see chapter 6), this, in turn, can be neutralized by more calcium. Nitrate ions are then available to be taken up by plant roots, so releasing the calcium to re-enter the lattice in an ionic merry-go-round. If the nitrate is not taken up, then calcium nitrate is likely to be leached out and lost into the ground water where it may eventually cause undesirable effects. So long as the soil contains a reserve of lime (calcium carbonate), loss of calcium through this process falls on this reserve, and not on the exchangeable calcium. However, after prolonged application of ammonium sulphate, and without liming, the soil may become acid, or 'sour'.

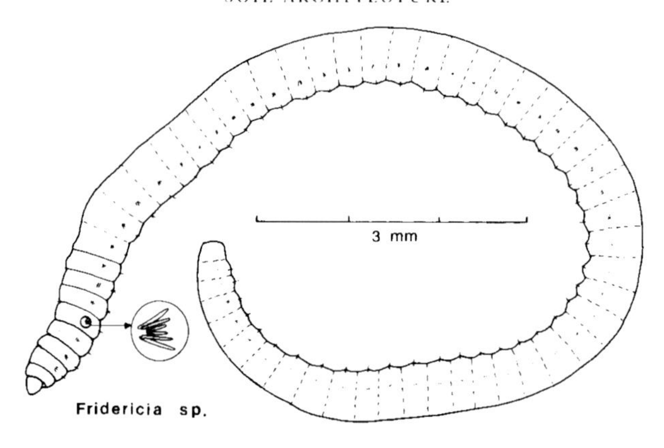

Fig. 5 An enchytraeid worm from woodland soil. They are like miniature earthworms but without pigment. Compare with Fig. 34. (Drawn by B.N.K.D.)

The aluminium itself may then start to be dissolved out of the clay lattice to form salts which are harmful to plants. This is one of the more subtle effects of 'acid rain' whose importance we are only now coming to understand.

Other fertilizers, such as sodium nitrate and potassium sulphate, act through the same processes of cation exchange, though with various minor differences. Illite clays contain potassium in the crystal lattice itself linking the silicon and aluminium sheets; though it is not readily exchangeable, it can act as a reservoir for potassium over a long period.

The power of attracting and exchanging ions also explains the way in which certain pesticides, in particular the herbicides diquat and paraquat (e.g. Weedol), are adsorbed and rendered inactive. Other less strongly ionized herbicides and insecticides are not bound by clays in this way. The chlorinated hydrocarbon (= organochlorine) insecticides (e.g. DDT) are not adsorbed by clays but bound by organic matter in a different way.

A more familiar aspect of soil chemistry is soil acidity for this is often expressed clearly in the vegetation. Acidity of aqueous solutions is measured by the concentration of free hydrogen ions. Pure water, H_2O, is considered to be neutral, and has one free H ion (and one free OH ion) to every 10 million water molecules (i.e. a concentration of 10^{-7}) in which the H ions are still firmly bound to the OH ions. A strong acid, such as hydrochloric acid HCl, 'dissociates' in solution much more readily, giving as many as one free H ion to every ten that are still attached (a concentration of 10^{-1}). On the other hand, caustic soda solution (1/10th 'normal'), produces very, very few free hydrogen ions – only one in 10 million million (10^{-13}). For convenience, this wide range of values is represented on a logarithmic scale as the negative power value of hydrogen ions. Thus the values given above for water, acid and alkali would be quoted as pH 7, pH 1 and pH 13 respectively – the abbreviation pH being derived from the original French expression 'puissance d'hydrogen'. (The feeble alkalinity of blood is more neatly expressed on this scale as pH 7.35 rather than as $10^{-7.35}$ or as 3.6×10^{-8}).

Gardeners are usually only concerned with the more extreme cases of soil acidity, typified by the ability or inability to grow heather and azaleas, or the need for lime to be sprinkled around brassicas. Farmers are aware of more subtle differences which are revealed in the growth of his crops or the productivity of his hill pastures. Such plant indicators, however, are just the above-ground symptoms of pervasive influences acting within the soil environment. Many animals with hard external skeletons, such as woodlice and snails, need calcium and therefore favour soils derived from basic rocks. Earthworms, too, need calcium for their digestive glands; most species react unfavourably to acid conditions, whereas their smaller relatives, the enchytraeids, are most numerous in slightly acid soils (Fig. 5). Soil microorganisms often flourish best under neutral conditions but many kinds of fungi grow in acidic situations; in very acid podzols (see next chapter), they are often the dominant members of the micro-flora.

The typical range of pH values in soils runs from about 3.5 to a little over 7.0. Soils in the range 3.5–4.5 (e.g. some heathlands) are considered extremely acid, soils with pH 4.6–5.5 as very acid, 5.6–6.5 as moderately acid, around 7.0 as neutral, and greater than this as calcareous or alkaline. Most arable soils are maintained by liming in the 5.5-7.0 range. At the extremes, colliery spoil containing the iron sulphide mineral, pyrites, can have a pH less than 3, while some industrial waste tips, such as those from the manufacture of sodium carbonate in Greater Manchester, can be extremely alkaline, as high as 12.7. The ecology and treatment of such chemically extreme, man-made, 'soils' are considered in chapter 10.

Clay minerals tend to stabilize, or buffer, the pH of soils by resisting changes in hydrogen ion concentration when small amounts of acid or alkali are added. If lime is added, free H ions are neutralized but the clay then releases more H ions to restore the original pH. Thus more lime is needed to produce the desired effect than would be expected: a soil containing 50% clay will need more than twice as much lime to raise the pH as a sandy soil with less than 10% clay. In natural conditions, sandy soils become acidic much more readily than clay soils because they lack this buffering capacity.

Soil pores, atmosphere and water

Lastly, in this survey of the soil environment, we return to the question of space within the mineral soil. Permeating the solid mass of mineral and organic particles there is a labyrinth of inter-connecting channels and pores differing greatly in size. These can be divided into capillaries or micropores less than 0.03mm (30 microns) in diameter, medium-sized mesopores, in the fine sand particle range of 0.03-0.1mm, and large macropores of more than 0.1mm or 100 microns. The capillaries serve as sites of activity for root hairs and micro-organisms – they are the laboratories of soil life. The larger pores act as service ducts allowing rapid movement of air, water and dissolved nutrients through the soil. They are also the living quarters for a wide range of true soil animals. Here, for example, we can find an assortment of protozoa, and a multitude of tiny nematodes (eelworms) which can scarcely be seen by the naked eye, small mites, white Protura and Thysanura (bristle-tails), blind and springless springtails, and multi-legged centipedes, pauropods and symphylids. These are mainly soft-bodied creatures, unable to burrow for themselves, but even the burrowing fauna, such as worms and slugs, millipedes and woodlice, the larvae of beetles and moths, and sundry others, make use of the larger pores and channels when they can.

Soil animals, plant roots, fungi and most bacteria need oxygen which can enter the soil easily through air-filled pores from the atmosphere above. Oxygen can diffuse through water-filled capillaries, but 10,000 times more slowly. In soil crumbs that are more than 3mm in diameter, all the available oxygen may be used up before it can reach the centre. Such conditions allow anaerobic bacteria to survive in most soils, including those like some *Clostridium* species that perform the important function of fixing nitrogen from soil air. Another group of microorganisms, the actinomycetes, can flourish in intermediate conditions of low oxygen tension. The activities of the micro-flora are described in chapters 6 and 9.

Soil air also contains carbon dioxide given off in respiration and by the decomposition of cellulose and other plant remains. Although the actual carbon dioxide concentration is low compared with that of oxygen or nitrogen, only about 0.5 per cent by volume, this is already ten to twenty times more than in atmospheric air. Small fluctuations in carbon dioxide concentrations therefore can have important direct and indirect effects on plant growth. Microbial activity can also produce ammonia, marsh gas (methane), hydrogen sulphide and other gases. Our sense of smell is very limited so we are unable to appreciate the galaxy of aromas that soils must have, but we can usually tell if a soil is sweet smelling or unpleasant, especially when anaerobic. Within the finer pores, the composition of the soil atmosphere may vary markedly around local microbial concentrations or other sites of biological activity such as plant roots.

While the overall pore space obviously sets a limit to the total volume of soil water and soil air, the balance between water and air is all important. Both are needed for the healthy growth of roots and for animal life. The optimum state, therefore, is when there is a sponge-like combination of fine pores to hold water, with larger pores to allow free drainage of excess water and access of air. This state can best be achieved when there is a good crumb structure, but soil texture is also very important as it determines the natural packing of mineral particles.

When all pore space is filled with water, the soil is saturated or waterlogged. Seasonal or permanent waterlogging is common in natural soils because of impeded drainage or high water table (chapter 2), and may severely affect soil animals and plants. When there is free drainage, some of this water drains away quickly by gravity, transporting gases, nutrients and other chemicals in solution down the soil column. The remaining water fully occupies the finer pore spaces of the soil, and at this stage its moisture status is said to be at 'field capacity'. This condition would be reached, for instance, after two or three dry days in winter following heavy rain; it could persist indefinitely in the absence of losses to the atmosphere.

This capillary-held water is the main source of water that is available to organisms. Protozoa, earthworms and plant root hairs alike depend on the uptake of water by osmosis through thin membranes. Plants, however, need a continuous supply to balance the water lost by transpiration from the leaves, and can quickly deplete the store of available water in the soil during the growing season; up to 25 tonnes a hectare (10 tons an acre) can be sucked up by a crop on a hot summer's day. When all the capillary water is used up, the soil reaches a 'permanent wilting point' when annual plants die, and microorganisms generally cease functioning. The root hairs and cell walls simply cannot withdraw any more water from the soil.

There is still some 'unavailable' water held hygroscopically as thin films around clay and humus particles. This water is invisible but can be measured by drying soil in an oven at 105°C, and measuring the loss in weight. Cotton and wool, for example, are also hygroscopic and can retain up to 8 per cent and 15 per cent respectively without appearing wet.

The concept of suction force provides a way of calibrating the water-holding properties of a soil. The actual amounts of water held by a soil at field capacity and at wilting point, and the rate at which the available water is released under increasing suctions, vary greatly with the soil texture. It is easy to show that a light sandy soil holds only about 3% of its volume of water at the wilting point, whereas a heavy clay still holds 20% or more. This is because there are more fine capillaries in the clay exerting a suction force on the water.

Suction force can be expressed in various ways. The simplest is to equate it with the pressure of a column of water; for example, field capacity in the laboratory is often taken to be equivalent to the pressure produced by a 50 cm column of water. If one thinks of tyre pressures, then this would be equivalent to a very soft tyre with a pressure of only 0.74 lb/sq in. or about 0.05 atmospheres. At the permanent wilting point, however, the pressure is equivalent to 15,000 cm of water, 220 lb/sq in. or 14.8 atmospheres. A commonly used scale to calibrate suction force is the logarithmic one introduced by R.K.Schofield in 1935. This he called the pF scale by analogy with the soil acidity (pH) scale described above. One thousand centimetres of water thus becomes pF 3, the permanent wilting point occurs at pF 4.18, and so on. To take account of metrification, modern texts often use Bar units (1 Bar = 1 million dynes/sq cm), and for all practical purposes 1 Bar = 1 atmosphere.

Moisture always tends to move from a wetter to a drier soil until the forces on both are equal, and the whole system is at the same 'moisture potential'. For the same percentage moisture content, a sandy soil will have a lower moisture potential than a clay soil, so that, if the two are brought into contact, moisture will be drawn out of the sand into the clay. In practice, capillary movement of water is so slow that plants cannot depend on this to provide adequate supplies. If a root uses up all the available water in its immediate vicinity it will have to grow 'in search of' more water. This slow capillary movement explains why, after a dry period, a shower of rain has such little penetrative effect; the surface layer of soil absorbs all the water, up to field capacity, and holds it against mere gravitational downward pull. Of course, where there is a high water table, as in a low lying meadow such as Wendlebury Meads (page 137), relatively little rain is needed to saturate the soil, and plants are liable to suffer from too little air in the pore spaces.

By applying a series of suction pressures to a range of different soil types, as defined in the triangular diagram shown earlier, one can map their relative water retention properties (Fig. 6). Sands, loams and clays differ markedly not only in total available water capacity but also in the proportions which are released under increasing suction pressures. For example, the clay loam used in that trial held only half as much available water altogether as the silt loam. Compared with the sandy clay loam, on the other hand, the total available water was nearly the same but the clay loam held much smaller proportions of this water at very low and very high suction pressures.

Many soil animals are rendered inactive long before the suction pressure

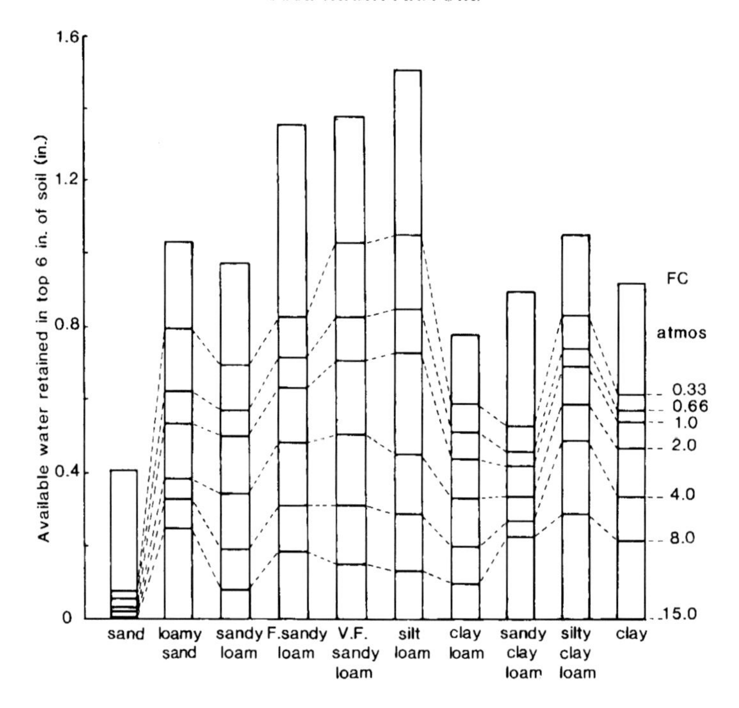

Fig. 6 Water retention characteristics for 10 different soil texture classes. Water is released at different rates when suction is increased (artificially or by the action of plant roots) from field capacity (FC) to permanent wilting point (15 atmospheres). This is shown by the horizontal bars on each column. (Adapted from P.J. Salter & J.B. Williams 1965.)

reaches the permanent wilting point. The interstitial fauna, such as protozoa, rotifers and eelworms, live in films of water in the soil pores, and it is known that most eelworms can only sustain activity when the pF is less than about 2.7 (about 0.5 atmospheres). This occurs when there is about 17 per cent water in the sandy soil shown in the figure but when there is as much as 50 per cent in an organic-rich soil. The availability of water in both cases is the same.

Agriculturalists like to relate the total available water content (AWC) to the rooting depth of a given crop. The top 900mm is conventionally taken as being potentially exploitable by the roots of seasonal crops. A typical AWC for a well drained sandy loam would be about 17% by volume or 150mm. Moving to the left along the bottom of the triangular diagram of soil textures, a loamy sand would have a low AWC of less than 110mm while a silt loam, at the bottom right hand corner of the diagram, would have more than 190mm per 900mm depth. With shallow soils overlying hard limestone, the rooting depth might be only half as much; in this case the available water content might be expressed as 75mm in 450mm.

Taken together with the average rainfall of a district, the total available water content gives a good measure of the cropping potential of the land. Thus in Eastern England, with rainfall of less than 625mm (24 in.), sugar-beet and cereals will yield twice as much, and potatoes three times as much, on soils with high AWC as on soils with low AWC. In areas with higher rainfall this discrepancy will be less. As a generalization, medium-textured soils have better characteristics for supplying water for plant growth than either coarser- or finer-textured soils. This is attributable to the fact that good soil structure is easiest to create in clay loams.

Farmers expend enormous efforts in cultivation, drainage and irrigation to produce the optimum balance of air and water in the soil. Optimum, that is, for the range of arable crops which society demands. The flora in Mesolithic times, before farmers started to cultivate the land, would have been perfectly well adapted to a quite different range of soil conditions. Life in the soil has had an unbroken history since the last glaciation, and even what we consider to be semi-natural communities in this country have themselves been modified by direct or indirect forms of land management over the centuries. Some of these management influences are considered in the next chapter, together with natural environment factors affecting the development and distribution of soil types.

2

The Variety of Soils in Britain

The building blocks of the soil world have been described in chapter 1 as components that, assembled in differing frameworks of solid particles and intervening pores, are the basis of a complex physical environment. This complexity is exploited by a diverse range of soil organisms for living space, moisture and food. In this chapter, we draw back from considering the fine details of soil composition, and view the wide variety of soil types that contribute to the character of different regions and localities in Britain. Many books give comprehensive accounts of soil-forming factors, soil distribution, alternative classifications, and the application of soil science to land use and management issues, on world, national, and local scales. Here we give just an outline of one way of looking at the formation and variety of British soils as a background to our main theme.

The traditional 'ecological' approach to the field study of soils is to consider vertical sections through this thin skin that covers most of the earth's surface. These sections are three-dimensional units, that is to say, they have both a surface area and a depth. One particularly important contribution to this approach was that of W. L. Kubiena, in his book on *The Soils of Europe*, mentioned in chapter 1.

When seen in section, the character of a soil can change laterally, either abruptly or gradually, as soil-forming conditions change. Additionally, there are changes in soil properties and appearance as one moves downwards, from the ground surface with its plant cover, to the unchanged parent material below. This parent material may be solid rock, like limestone, or a relatively soft geological deposit, like London Clay, alluvium or wind-blown sand, or a man-made deposit such as colliery waste.

In many soils it is immediately obvious that the vertically changing cross-section of the soil unit – the 'soil profile' – consists of a sequence of distinct layers, or 'horizons'. In others, although changes are taking place down the profile, they are visually less conspicuous. Whether soil horizons are sharply distinct or not, they differ in properties such as their moisture regimes, particle size composition, handling consistency, organic matter content and other chemical properties. The consequences of these differences are frequently revealed as changes in colour, or in the feel of the soil when it is handled. Although equivalent horizon thicknesses vary, their sequences down profiles are seen to fall into consistent arrangements that are repeated from place to place.

A common way of studying, describing and comparing soils is to dig soil profile pits. These may need to be two or three feet deep to expose the average soil profile in Britain, and are appropriate for serious survey, teaching

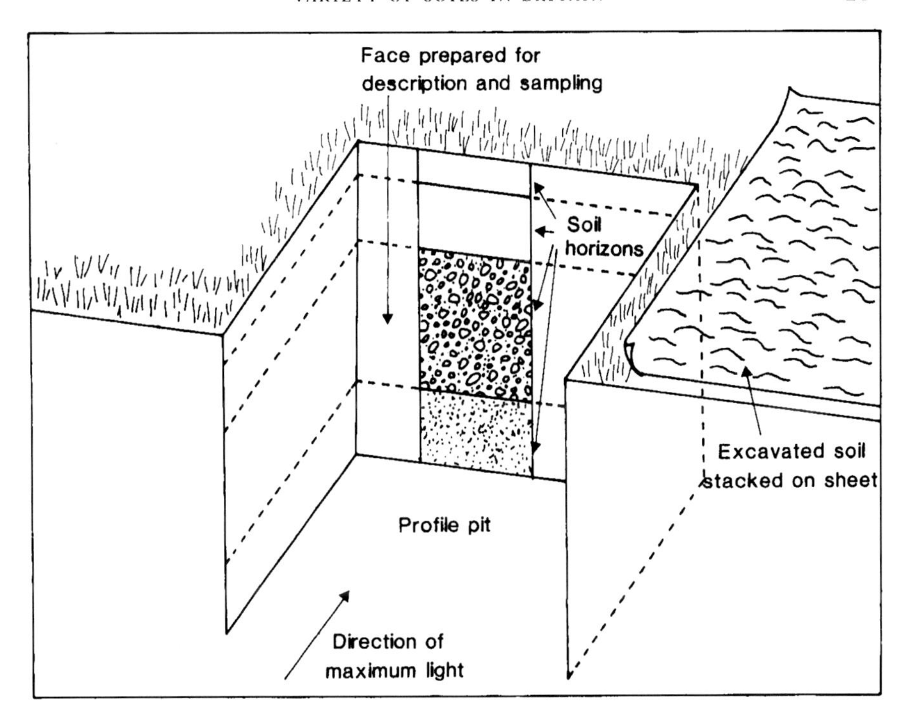

Fig. 7 Sketch of a soil profile pit. (From D.F. Ball 1986.)

and demonstration purposes; they are unlikely to be popular with landowners if their purpose is merely to satisfy individual curiosity! Figure 7 sketches a typical pit. Alternatively, for quick examination of soils from point to point and with minimal disturbance, a screw-thread auger (Fig. 8) is used in professional studies to withdraw successive small depth increments from all but the driest or stoniest soils. Often, the easiest way to appreciate the lateral and vertical variations in soils is to study suitable natural or artificial sections. Freshly exposed by road cuttings, in quarries or on construction sites, these sections can give a more comprehensive three-dimensional view than any other option does. However, their occurrence is of course random with respect to soil patterns, or to any particular interest, rather than being chosen in accord with some statistical or methodical sampling plan.

The repetition of similar horizon sequences in different profiles provides a kind of bar-code that can be an identifier of a soil class. Using such sequences for classification, soils of particular profile types can be considered as natural units. These, whatever their location, are assumed to have had, and usually still have, their soil-forming ('pedogenic') factors in common. It is this pedological view of soils as natural bodies, created through consistent and orderly interactions with formative environmental factors, that will be emphasized here.

Before turning to the influences producing different soils, two points need to be made. First, it must be appreciated that there are other approaches to soil classification. Valid single-purpose or general classifications, and soil

Fig. 8 Sampling soil profiles with an auger; W.M. Williams on Snowdon. (Photograph D.F.B.)

distribution maps, can be produced solely from observed data. Such systems do not require reference to, or presumptions about, relationships between soil units and environmental controls. The data can be individual soil properties, for example, surface soil acidity, texture, or organic matter content. Soils are then just grouped into classes based on ranges of the property used. This is often the course used to define soils in biological studies. Alternatively, the data can be compiled for many variables, determined at more than one depth. These data can be analysed by computer classifications that allocate each recorded individual soil into a statistically determined group. Such methods simply accept that the soil categories are valid if they are found to be useful for a particular purpose. There are different opinions among protagonists of the traditional and numeric schools concerning the principles and practice of soil classification. However, it is generally accepted that both approaches have their value. The numerical approach can be tailored to specific management issues more readily than can the general-purpose approach of the environmentally based schemes. These latter, on the other hand, with their linked galleries of classes at different levels of division, can round out the geographic picture; they can show relationships between the physical and biological worlds, and indicate the role of soils as a natural resource. The traditional classifications have survived criticism and remain continuously in active development and use, ever since their origins a century or so ago in the work of pioneering scientists, notably the Russian, V. V. Dokuchaev.

The second point is that, whatever type of classification is favoured, at whatever level of detail, actual landscapes are usually more complex and less orderly than might be assumed from the idealized world of classification. The best any soil map can do is to show, with greater or less precision, the probability of a specified soil class existing at any particular spot. Within a single soil class, significant variations in chemical and other properties usually occur between representative profiles as a whole, or between equivalent horizons within them. Management and seasonal climatic differences extend the inherent variability that occurs naturally over very short distances, even within an uncultivated and morphologically uniform soil. Over Britain as a whole, broad geographic gradients create zones with characteristic groups of soils. Within these zones, however, much local diversity can be superimposed. From the resulting complex soil picture, even the most efficient classifications have to use arbitrary divisions in distinguishing particular types. Plant ecologists (phytosociologists) have grappled with similar problems in classifying plant communities.

Thus, in any classification, a given profile class or horizon type does not strictly define its population of soil organisms, in species or numbers. These classes have a broad two-way relationship with the character of the soil fauna as a whole, but this is also strongly affected by the particular vegetation cover and its consequent litter fall, and by past and present land management. The versatility with which some organisms adapt to different environments means that they can occur widely under differing soil types or land uses. Chapter 10, for example, describes the recurrence of the red worm *Lumbricus rubellus* on a variety of industrial spoil materials and landfill as well as in natural grassland. Other species have specific controls imposed on them by soil properties, singly or in combination. Such interactions between soils and their fauna are particularly clear under semi-natural plant communities. Some examples are considered in chapter 7.

Soil-forming factors

The key environmental controls on soil formation are those of climate, parent material, landform (i.e. relief or physiography), biological factors (plants, animals and man); and time over which these factors operate. Of course, the different factors act simultaneously, and their individual importance varies in different situations.

The broad geographic division of Britain into regional soil zones results from the wide climatic range across the country. The significance of climate is also dominant on continental and world scales, and was the key to the distribution of major soil zones that was recognized in Russia by Dokuchaev and his fellow scientists. Rainfall and temperature combine to influence the speed and type of weathering of rocks; the directions and rates of transfer of plant nutrients and other mobile chemical constituents through the soil; the form in which organic matter accumulates; the natural flora and fauna; and the options for land use.

In this broad geographic zonation, a cooler and wetter 'highland' zone in the west and north contrasts with the warmer, drier 'lowland' zone of the south and east. These zonal terms reflect a general environmental idea, rather than implying a simple boundary at a particular altitude. The dominant rain-bearing winds in Britain are from the south-west. Steep increases in rainfall take place from the western seaboards to the central highland

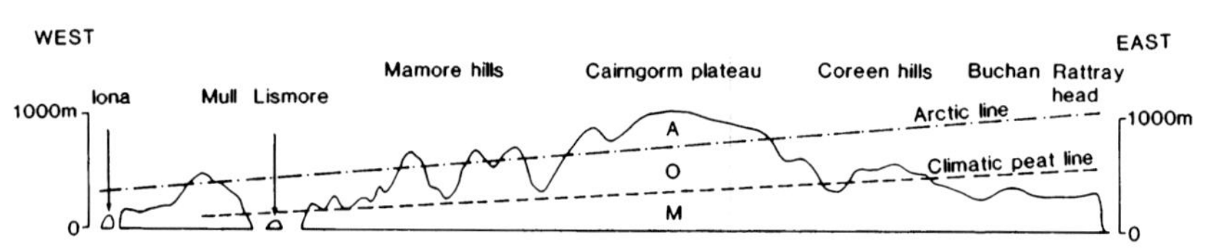

Fig. 9 Above, A wind-scoured, boulder-covered summit plateau in the 'arctic' soil zone; Cairngorm Mountains, Scotland. (Photograph D. F. B.). *Below,* Soil zones defined by climatic trend lines on a transect across Scotland. (Adapted from J.S. Bibby 1986.)

spines of Scotland, northern England and Wales. Lower rainfalls occur at similar altitudes in the rain-shadow areas to the east, compared to the exposed west. In North Wales, for example, annual average rainfall is about 950 millimetres at sea level at Bangor on the coast. It rapidly reaches 1500 millimetres 5 kilometres inland at 150 metres altitude, and is 3000 millimetres or more at around 1000 metres altitude (3281 feet), only some 12 kilometres to the southeast on the Snowdonia peaks. In the rain-shadow area to the east of these peaks, 1500 millimetres annual rainfall is hardly reached on the highest parts of the Denbigh Moors at around 500 metres altitude.

Temperatures fall with increasing altitude. Although this drop (the 'lapse-rate') varies seasonally at one place, and from place to place, it averages close to 0.6°C for each 100 metres rise in altitude. For example again, in the Pennines the summit area of Kinder Scout, at 630 metres, has an average temperature some 3°C lower than at Stockport, only 18 kilometres to the west but at an altitude of 100 metres.

The climatic influences on the soils of the highland zone are accentuated by the prevalence of hard, less easily weathered rock types in the west and north. Within the highland zone, there are distinct vertical zones that result

Fig. 10 'Patterned ground', produced by frost and wind action in the 'arctic' soil zone, can occur at low altitudes in northern Britain, as here on Hoy, Orkney (Photograph D.F.B.)

from the regional relationships between altitude, rainfall and temperature. Figure 9 illustrates this by a schematic cross section through Scotland showing three soil zones, divided by approximate climatic boundaries that fall steadily in altitude from east to west. An upper 'arctic' zone starts above about 400 metres in the west, but not until above some 700 metres in the central Grampians. (It would be much higher than the actual land surface in the east). Within this zone, bare rock surfaces are frequent together with areas of immature shallow soils, often mobile and frost-heaved. These are seen on the summits of the Cairngorms at around 900 metres (Fig. 9) where soil conditions have much in common with those above 2700 metres in the Alps. Locally in this 'arctic-alpine' soil zone in Britain, one finds 'patterned ground' terrain (Fig. 10) and soils characteristic of high alpine and sub-arctic locations. North of the transect position in Figure 9, such features are found on hills as low as 275 metres altitude on suitable rocks in Orkney and Shetland. Well to the south, in the Lake District and North Wales, these active arctic-alpine features occur only in limited areas above about 900 metres, although relic frost-action landforms, from earlier climatic conditions, are still recognizable at lower altitudes.

The lower, 'climatic peat', zone of Figure 9 is at about 100 metres altitude at the western end of the transect, and rises to about 400 metres on the eastern hill areas. To the north, in the Western Isles and on the north-western Scottish coasts, it comes down to sea level. It is characterized by moorland vegetation on moderately drained and poorly drained soils with peaty surface horizons, and on associated deep peats, the latter being dominant in many Scottish and Pennine areas. In North Wales, a similar transect from west to east shows that such moorland soils occur above an average of 350 metres in the wetter west, and above 550 metres in the highest hills of the drier east.

Within these vertical zones, there are of course major soil contrasts brought about locally by different parent materials, and by different landform features such as locations on hill crests, upper and lower slopes, or in basin sites. Some such soil contrasts from Snowdon are discussed in chapter 7.

Between the broad climatic zones of hill and lowland there is an intermediate belt, fringing the moorlands. This is narrow in some areas, and geographically more extensive in others, as in Wales. The freely drained soils of this 'upland margin' zone combine something of the character of the principal soils of its two neighbouring climatic zones, as described later in this chapter.

In the extensive lowland zone that dominates southern Britain, and also extends up the eastern side of Scotland, regional climatic differences have little direct effect on soils, though they do affect the possible crop options and their yields. Rainfall and temperature variations from year to year at one place can be greater than differences in long-term averages between places. The contrasts between soil classes in the lowlands, which are very great, are due mainly to the influence of parent materials, and, within a single parent material, to landform.

Contrasts in parent material range between almost pure calcium carbonate rocks such as the chalk, which in most locations carry shallow calcareous soils, and virtually pure silica sands and sandstones, on which highly acid soils develop. Although many soils form directly from parent bedrock at shallow depth, others are developed on transported geological deposits.

It may be thought that a geological map would show what parent material is present at any particular site. However, such maps display either the stratigraphic age of the solid rocks, or the mode of origin of the overlying alluvial or drift deposits, rather than their composition, which is what influences soil properties. Limestone bedrock shown on the geological map may, for instance, be overlain by shallow silty glacial drift derived from non-calcareous shales outcropping elsewhere. Additionally, within one rock age class or map unit there may well be considerable variation in rock chemistry and other relevant properties. Bedded sandstones, shales, and limestones may all occur in rapid succession at one place in a single rock unit, while rock of the same age-based unit can be limestone in one area and sandstone in another. Thus, although geological maps give a general picture of the potential source materials for soil formation, only field examination of soils themselves, and possibly laboratory investigation also, can confirm their parent material origins.

Among landform influences, steepness of slope and position on the slope have the most important effects on soils. Together with soil texture characteristics described in the previous chapter, they control drainage as illustrated in Figure 11. On clay-rich, poorly permeable materials, waterlogging persists near the surface on both level and gently sloping ground; poorly drained soils commonly occur on crests and lower parts of slopes while freely drained soils can be restricted to quite steep slopes. On lighter-textured (sandy) parent materials, however, freely drained soils predominate. The water that has moved through the soil flows downslope in the underlying parent material, and can create a high water table at the base of the slope and in valley or basin sites. A sequence of soil types, formed on the same parent material but differing in drainage according to their position on a slope, has been called a soil, or drainage, 'catena'. Figure 11 shows two

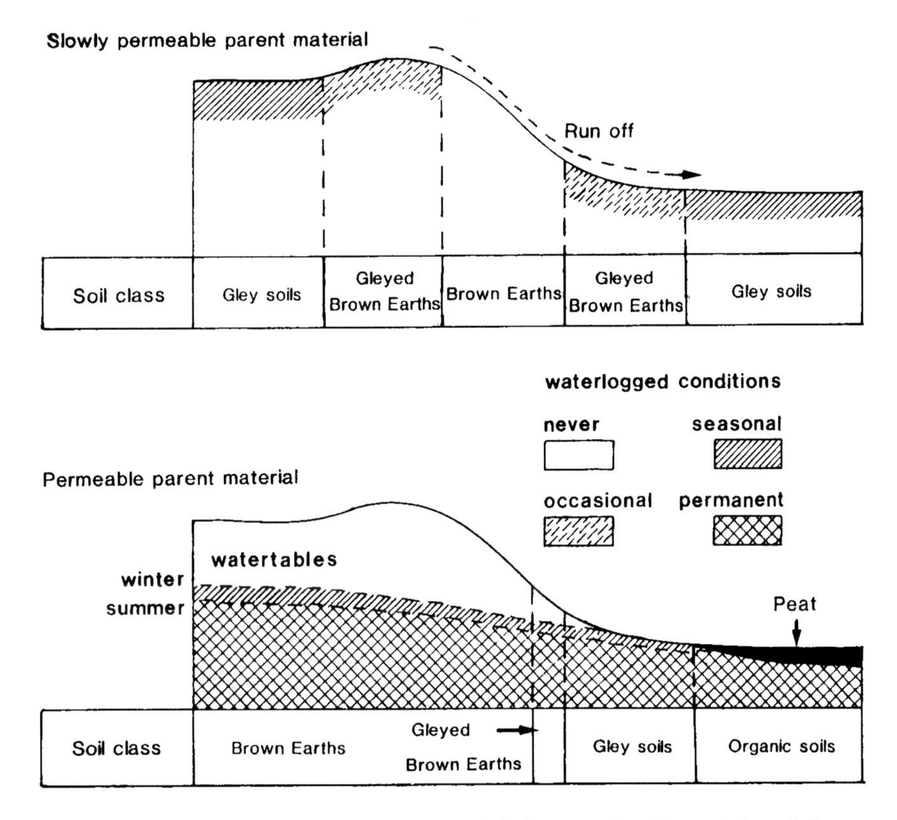

Fig. 11 The relationship between soil class and drainage to landform (Adapted from C.C. Rudeforth and others 1984.)

variants of such catenas, which are widespread features of soil variation on a local scale.

Among biological factors, some direct effects of vegetation and organisms on soil profiles are considered in chapter 7. In an intensively settled and agriculturally developed country like Britain, man has had an important effect on the course of soil development everywhere at some time or another, and has remained a major influence through historic time in lowland Britain. The evidence of soil change on abandoned arable land in Wiltshire, also described in chapter 7, is one example of man's influence. Another is the way drainage, and, to a lesser extent, irrigation, rapidly alters the moisture regime in a soil. In such circumstances, profile characteristics of a formerly poorly drained soil can persist, even though the conditions that produced these characteristics have changed. The biological possibilities for soil life will also have changed, while the inherent soil structural framework remains.

Erosion is an outcome of historic and recent activity by man. Although there is a tendency to think of soil erosion as a problem only for other parts of the world, this is not so, and some recent aspects of soil erosion in Britain are considered in chapter 9. In low rainfall areas on light soils, windblow at the wrong time can remove whole surface layers of exposed soil, complete with seeds and seedlings. In the hills, the natural erosive forces of wind, rain, frost, and steep slopes, are added to by heavy grazing, by burning, and by access roads and other forestry and recreational activities. Together, these produce landslips, gullying, the wash of eroded material downslope, the destruction or burial of original soil profiles, and the supply of sediment to rivers, lakes and reservoirs. In the geological long term, of course, erosion from the land, and sedimentation in the seas, is part of the cycle of creation of new rocks and continents. On a human time-scale, such a philosophical view of the world ecosystem is more difficult to accept.

The pedogenic factors of climate, landform, and biological influences have had, in general, some 10-14,000 years of post-Glacial time to achieve their effects on parent materials in Britain. During such a period, there have been long spells with differing climates, and, in the latter half of this period, the progressive effects of man as settler and cultivator. These natural and human factors have both affected trends in soil development. Soil scientists, archaeologists, and palaeobotanists (specialists in pollen analysis) have been able to understand some of these effects through time, from evidence preserved in some soils and peats. The onset of peat growth on the Pennines between 7500 and 5000 years ago is one finding. Another is the increased leaching of lime, and consequent soil acidification, that apparently affected some upland and hill areas in the Bronze Age around 3500 years ago. As a result of deforestation for agriculture during the relatively warm and dry continental climate, the rather finely poised stability of the existing soils was disturbed. The initial impetus caused by vegetation changes was accentuated by later climatic deterioration to a cool, wet, oceanic climate. This brought about extensive podzolic soil development (see below), and the acceleration and spread of the trend to more extensive moorland soils.

It is sometimes possible to follow profile development more directly over quite short time scales. An example comes from a sand dune system in Norfolk. Here, a series of linear dune ridges has formed, and is still actively developing successively to seaward, with new ridges protecting the older, inland, dunes from fresh supplies of wind-blown sand. Thin continuous mull humus horizons become distinct within a few decades under dune grassland on the youngest stabilized ('fixed') dunes. After 15 years of stability, these horizons were 2 centimetres in depth; after 30 years 5 centimetres; and after 70 years 9 centimetres.

The oldest inland dune had been planted some 90 years before with Austrian pine *Pinus nigra*. This had become closed canopy woodland, and self-sown pines from here have colonized the adjacent dune ridges. The effect of acidifying needle litter on the planted dune has redirected soil development. The grassland mull humus horizon on these oldest ridges has been replaced by a mor horizon some 7 centimetres deep. Leachates from the pine needle litter have accelerated the rate of destruction of the original shell particles in the sand, removing the dissolved calcium in drainage water. Together with leaching of the limited initial iron content, the result is a thin, but well defined, bleached horizon below the organic layer. As noted in chapter 1, sandy soils lack the capacity to retain elements like calcium, and can therefore become acidic quite quickly. After 30 years under dune grassland, the pH of the sand in the top 15 centimetres was around 6.8, whereas under the oldest pines it had decreased to 5.0.

Trends in soil development and associated faunal changes over quite short periods have also been observed in colliery spoil and other mineral workings. These are considered in chapter 10.

Soil classification

We said earlier in this chapter that horizon sequences in profiles provide a 'bar-code' that can be used to distinguish soil classes. In this code, definitive symbols are allocated to particular types of horizon. These types are identified by position within the profile and by observable properties, supplemented in more detailed systems by laboratory analysis. Soil classes at the most broadly defined level – the 'major soil group' – have specific horizon sequences. Widely used conventions for horizons include variations on the following symbols, which we apply later in an outline of some widespread British soil groups. In comprehensive schemes, a much longer list of such symbols is used.

O		horizons dominated by organic matter; subdivisions include -
	L	recently deposited plant litter,
	F	partly decomposed organic matter that retains some original plant structures,
	H	decomposed organic matter, no longer retaining original plant structures. In other usages, L, F and H symbols are used for organic horizons in freely drained (aerobic) conditions: O for peaty organic matter formed under wet (anaerobic) conditions.
A		mineral horizons of maximum biological activity, at or near the ground surface, in which organic matter is intimately incorporated; subdivisions include -
	A_h	an A horizon with notably high organic matter, but not sufficient to be an O horizon
	A_{Ca}	the Ca subscript is used for horizons containing calcium carbonate, as rock fragments or secondary concretions.
E		mineral horizons below O and/or A horizons from which iron and aluminium and/or clay particles have been mobilized and transported to lower horizons, or out of the profile in drainage water; subdivisions include -
	E_a	horizon from which iron and aluminium have been lost.
B		mid-profile mineral horizons, modified by physical, chemical and/or biological activity, so that they differ from horizons above, and from the parent material below; subdivisions include -
	B_s	B horizon notably enriched in mobilized iron and aluminium compounds, released in situ or transported from above
	B_h	B horizon enriched in transported humus.
C		parent material substantially unmodified by soil-forming processes.
G/g		these are used as subscript symbols for A, E, B, and C horizons which show strong/moderate evidence of the effects of long-term/seasonal waterlogging.

In uncertain or transitional cases, intergrade horizon symbols (e.g. BC) are often used in profile descriptions.

There are many variants of international and national soil classifications, which, over the years, have been based on the work of many scientists of various nationalities. Among current international schemes, that of the Food and Agriculture Organisation of the United Nations (FAO) is in use for soil maps on continental and world scales. The United States Department of Agriculture (USDA) 'Soil Taxonomy' is a complex system of wide actual and potential application. It has been developed and refined over many years, in successive 'approximations', to its present form, which uses its own, self-contained and specialized, terminology. You will sometimes see class names from these systems cross-referenced in papers describing soils by national systems.

In Britain, there are currently two 'official' systems in use which, unfortunately, differ in their terminology, and in the application of similar technical terms. Both, however, are intended to be fully comprehensive for their respective areas. They have been established by the Soil Survey of England and Wales (now the Soil Survey and Land Research Centre, based at Silsoe, Bedfordshire) and the Soil Survey of Scotland (of the Macaulay Land Use Research Institute, Aberdeen). The 1:250,000 scale maps and accompanying 'Bulletins' of the former, which give a complete cover of England and Wales, use a classification that starts with 10 'major soil groups'. These are divided into 34 'groups', 67 'sub-groups', and finally some 700 'soil series'. These last are soils of a particular sub-group formed from a specified parent material, defined by soil texture (particle size composition) and source rock types. The 296 'soil association' map units at the regional scale are complexes of named dominant and subordinate series. The criteria for differentiating these series, and an index to them, requires a publication of some 150 pages.

The Scottish 1:250,000 scale maps and accompanying 'Handbooks' begin with 5 'soil divisions' which are progressively split into 12 'major soil groups', 36 'major soil sub-groups', 110 'soil associations' and 580 mapping units. These mapping units are comparable in concept to 'soil associations', as the term is used in the England and Wales survey, but they cannot be directly related to each other.

Variations on these traditional classifications have been developed over the past 60 years and can cause confusion. Some approaches use simple schemes, others involve much more detail. The former have the advantage of presenting a broad picture but the disadvantage that a high proportion of real-world soils fall between their classes. One such classification, derived from existing conventions, was suggested by one of us (D.F.B.) elsewhere as an appropriate baseline for broad ecological/soil correlations. It used seven major soil groups divided into 32 sub-groups, but for present purposes only 12 of these are emphasized here as giving a suitable background for appreciating soil diversity in relation to biological themes. Some soil names are self-explanatory in English while others (rendzina, gley, podzol) are derived from languages of other countries, particularly Russia, where their distinct characters were first recognized and described. Their typical horizon sequences are listed first, followed by a brief outline of the nature of each soil. Conventionally, and throughout this book, these soil names are treated as proper nouns, i.e. with capital letters.

Raw Mineral Soils	soils lacking continuous development of any horizon above little altered C parent horizon.

Peat Rankers	non-calcareous soils with O,C profiles, i.e. with peat resting directly on parent material.
Brown Rankers	non-calcareous soils with A,C or A_h,C profiles.
Rendzinas	calcareous soils with A_{Ca}, C_{Ca} or A_{hCa}, C_{Ca} profiles.
Brown Earths	soils with A, B, C profiles.
Gleyed Brown Earths	soils with A, B_g, C_g profiles.
Gleys	soils with A_g, B_g, C_g or C_g profiles.
Peaty Gleys	soils with O, B_g, C_g profiles.
Brown Podzolic Soils	soils with A or A_h, B_s, C or C_g profiles.
Podzols	soils with O_h or A_h, E_a, B_h and/or B_s, C profiles.
Peaty Podzols	soils with O, E_{ag}, B_s, C or C_g profiles.
Organic Soils	soils with O horizons only.

Raw Mineral Soils, also called Immature Soils, include materials such as recently deposited wind-blown sands, or emergent marshland silts. The initial establishment of vegetation on these new land surfaces is followed by development of soil horizons as the surfaces become stabilized. The immature soil can develop further from this stage into other fully developed soil types. If the transient nature of surface stability persists, then developed profiles can become covered by a further supply of new material. Dunes, in fact, often show several buried profiles superimposed on each other, representing cycles of alternating stability and instability. In the marshland case, their silts are likely to have been derived from different horizons in other soils, before their transport in rivers and estuaries to become fresh marshland. As they become cut off, naturally or artificially, from further surface cover in their new situation, they effectively become a new parent material.

Rankers are shallow soils with simple profiles, in which mineral or peaty (A or O) horizons directly overlie parent rock (Fig. 12). Peat Rankers, such as

Fig. 12 A Brown Ranker soil on shale rock, near Conway, North Wales. (Photograph D.F.B.)

Fig. 13 Brown Earth soil on volcanic ash, Llydaw, Snowdonia. (Photograph D.F.B.)

those on Snowdon described in chapter 7, typically have between 15-30 centimetres of peaty material with a subordinate mineral component. Although their profiles allow water to drain very freely, these soils actually remain in a wet condition for much of the year under the high rainfall and low temperature conditions in which they form. Brown Rankers are similar shallow soils but with mineral A horizons, developing generally in moderate to low rainfall areas. They have a low water-holding capacity, and also drain very freely, so are susceptible to seasonal drought. They are associated with broken relief and rock outcrops, often in conjunction with Brown Earth variants on glacial drifts, or with Brown Podzolic soils in the upland margins. A typical profile on shale rock is seen in Figure 12.

Rendzinas are also shallow soils, but formed over calcareous parent materials such as chalk or limestone. By tradition, lime- rich soils have been conventionally separated as a distinct group because of the conspicuous ecological effects of their soil chemistry, even though their profiles are of similar character to the acidic Brown Rankers or Brown Earths. Chapter 7 describes the influence of time on soil properties in Rendzinas on chalk grasslands where cultivation had ceased some 30 years (Plate 2) and 130 years previously. Over this period, organic matter content increased from 8 to 22 per cent in the top 10 centimetres of the soil, giving a darker-coloured A_h horizon in the older soil.

Brown Earths (Fig. 13) are freely drained soils characterized by a deep, well mixed mull humus surface horizon. Weathering in these soils involves a relatively slow, low-intensity, chemical alteration of the primary mineral constituents. The iron and aluminium hydrous oxides that are released through this process remain *in situ*, to accumulate as stable coatings on mineral grains and in soil structural units. With moderately acid pH values in uncultivated conditions, the horizons of Brown Earths merge as they change character gradually down the profile. The soil illustrated in Figure 13 was formed over volcanic ash in Snowdonia. Soils of this group, on a wide range of parent materials, are important in the lower rainfall regions of Britain. Their free drainage and ease of cultivation make them the most generally favoured for arable cropping.

In association with these freely drained soils, imperfectly drained **Gleyed Brown Earths** occur on gentle slopes that receive run-off or groundwater from above, when soil texture and structure cause a moderate check to through drainage. The Bg horizons are of duller grey-brown colour, with fine rust-like mottling along root channels brought about by alternation of anaerobic and aerobic conditions.

In aerobic conditions the iron-oxidizing bacterium *Thiobacillus ferrooxidans* converts ferrous iron compounds to ferric compounds (see chapter 6, page 118). It is the ferric oxides that give the brown colour to freely drained Brown Earths. Gleying is a process that occurs during anaerobic conditions caused by saturation of soil pores by water. *T. ferrooxidans* is unable to operate and the soil has a grey colour produced by the ferrous iron compounds. The resulting **Gleys** are found, therefore, where there is poor or very poor drainage, both in lowland and moorland situations, often as one member of a catena of soil types on sloping ground, as mentioned earlier.

Land form, regional climate and soil texture (the prominence of clay, and to a lesser extent, silt) all affect the relative frequency of Gleys, Gleyed Brown Earths and Brown Earths in lowland soil associations. The agricultural system of 'ridge-and-furrow' created surface relief partly in order to improve the drainage of such gleyed soils. The ecological effects of this system are discussed in chapter 7. Although, today, Gleys are mainly used for grassland farming, modern sub-soil drainage can maintain the water-table lower than it would be naturally, and thus allow a wider range of agricultural uses.

Peaty Gleys are prominent moorland soils in the uplands and hills. They fall between rather better drained Peaty Podzols (see below), and the more or less standing water conditions of deep peat Organic Soils. Their wet, generally very acid, O horizons overlie strongly gleyed, pale grey mineral horizons.

Podzolization is a weathering and leaching process. It can be thought of as taking place in two stages. The first is release of iron and aluminium compounds, the second their transport and re- deposition within the profile. As mentioned above, iron and aluminium oxides released by mineral weathering in Brown Earths are retained in the A and B horizons as persistent coatings around soil particles. Where there is a very acid parent material, and/or strongly acidifying litter (e.g. pine needles), perhaps also combined with high rainfall, the mull humus of Brown Earths is replaced by superficial accumulations of moder or mor humus. As in the sand dune example described earlier, organic acids leaching from these surface horizons attack the soil minerals more strongly and carry dissolved salts further down the profile.

Brown Podzolic Soils are geographically important in the marginal upland zone between the zones dominated by Brown Earths and mineral Gleys, or by peaty moorland soils. Pedologically (in profile character), they are also transitional, fitting between these Brown Earths and the strongly developed Podzols or Peaty Podzols that are described below. Laboratory analyses show them to have a higher content of 'free' iron oxides than the equivalent Brown Earths, that is to say, this iron is more readily extractable by chemical means. However, this released iron is largely retained within the profile of Brown Podzolic Soils, as it is in the Brown Earths. The delicate balance between vegetation, soil fauna and profile character in these transitional podzolic soils is illustrated by the Scottish moorland/birch woodland cycle discussed in chapter 7. This balance can be disturbed by planting conifers which produce a deep, acid mor humus. A more strongly developed podzol profile then quickly develops, with a thin E_a horizon forming beneath the mor.

In strongly developed, freely drained **Podzol** profiles, the iron and aluminium are moved down the profile as organic salts (complexes). Their removal, particularly that of the released iron, leaves a pale greyish-white mineral skeleton dominated by the resistant silica mineral, quartz. This is the E_a horizon. The transported material is deposited in underlying B_h and B_s horizons, as microbial and fungal action destroys the organic part of the transporting complex. The necessary conditions for such well-developed Humus-Iron Podzols include very acidic, free-draining parent materials, such as quartzite rock scree in the Highlands, or sands and gravels in low rainfall areas. Such soils are an essential element in the ecology of English heathlands. The typical profile in Plate 3 is from the Vale of York.

In the sequence of moorland soils, we mentioned earlier that **Peaty Podzols** are relatively well drained. They nevertheless have an O horizon characteristic of wet, anaerobic, surface conditions due to the high rainfall in these areas. Beneath the O horizon is a leached, grey-coloured E_{ag} horizon which results partly from podzolic release and transport of iron and aluminium, and partly from the added effects of gleying. For this reason, such soils have been called Peaty Gley Podzols. The gleying is caused by the sponge effect of the persistently wet O horizon above, and by impeded drainage below. The latter is the result of a thin, hard band, or 'iron-pan', produced at the base of the E_{ag} horizon. Beneath the iron-pan is typically a more diffuse bright reddish-brown B_s horizon, enriched with transported iron compounds, in a moderately well-drained position in the profile.

The last soil class in the short list summarized here is that of **Organic Soils**. Sub-groups among these peaty soils are based on the nature of the underlying, or surrounding, geological material, which affects water chemistry. Variations also depend on the type of vegetation that produces the peat, and on the landform in which the soils occur – for example, valley, basin or plateau. Lowland peats can be used for arable farming after they have been drained, as in the Cambridgeshire Fenland where the peat itself then gradually disappears, or they can be more directly removed by digging for horticultural use, as in the Somerset levels.

Blanket and hill peat occurs widely, especially in northern Scotland and Ireland. In local economy, this can be hand cut for fuel. Commercially, it is intensively exploited in parts of Ireland, and for the distillery industry in the Western Isles. Pollution and erosion are further destructive forces. In both the lowlands and uplands, the loss of semi-natural peat habitats can be a source of dispute with conservation interests.

In concluding this look at some important soils in Britain, it is worth stressing the value of minimally disturbed soil types for future study. Nature conservation sites, such as National Nature Reserves and Sites of Special Scientific Interest, are primarily selected for their above-ground biological significance, but they can, incidentally, protect soils representative of a wide range of habitats and terrain. Our ability to re-create plant communities of desired ecosystems, such as floristically rich meadows, is increasing. We can certainly restore some of their characteristic vegetation, and perhaps attract certain elements of the associated fauna. However, re-creation of the historic 'archive' and 'benchmark' record that is preserved in soil profiles is impossible.

3

Roots

Virtually all flowering plants and ferns have roots, except for a few floating plants, such as the duckweed *Wolffia arrhiza*, and some parasitic plants, such as dodder *Cuscuta*, though these have structures which penetrate their hosts and act rather in the same way as roots. Some plants have very limited root systems, especially the non-green orchids such as bird's- nest orchid *Neottia* or ghost orchid *Epipogium*, though they are really a special case of parasitism, as we shall see.

As with so many aspects of biology, it is much easier to recognize a root than to define it. The most familiar type is the root that develops from a germinating seed and branches more or less profusely, but many plants have adventitious roots that arise in dense clusters from the base of the stem, or even from leaves. The long, little-branched roots of creeping buttercup *Ranunculus repens* or arrow-head *Sagittaria sagittifolia*, the roots that spring from the base of a willow branch kept in water, and even the roots that form on the edge of a leaf of cuckoo flower *Cardamine pratensis* if it is pressed onto moist earth, are all adventitious roots. Figure 14 illustrates both types of root.

A seed of a typical herbaceous plant, such as clover or lettuce, has a preformed root already present. It is usually this root that first breaks through the seed coat and initiates germination. The young root has a growing point or meristem at its tip, protected by a cap of loosely adhering cells. The meristem continually produces new cells, and immediately behind the tip these start to elongate, pushing the tip through the soil. In the process, cells of the cap are broken off by soil particles, but again they are replaced by the meristem. In some species, such as apple, the cap has a distinct point which may improve its ability to penetrate the soil, but in others it is more rounded.

All the growth of a root system, except for radial expansion of the older roots, takes place in this tiny region, a few millimetres long, just behind the root tip. The rest of the root system is therefore unable to change its position and so a good distribution of roots, with respect to the pattern of water and nutrients in soil and to other roots, depends on the ability of the root tip to grow in the best direction and at the best rate. The root tip can sense the conditions of the soil it is growing through and can alter direction, for example to avoid dry or toxic soil, rather like a miniature sense organ. The other important piece of root behaviour in this respect is branching. Behind the zones of cell production and elongation, there is first a zone in which the root produces hairs, tiny excrescences from the surface cells, most often only 1 or 2 millimetres long, which help to maintain contact between root and soil and to increase the surface area for the absorption of nutrients. Further back, typically about 2-5 cm from the tip, branches appear. These grow out from the deeper tissues of the root, at first at right angles to the parent, but then changing angle progressively as they get further from its influence. These

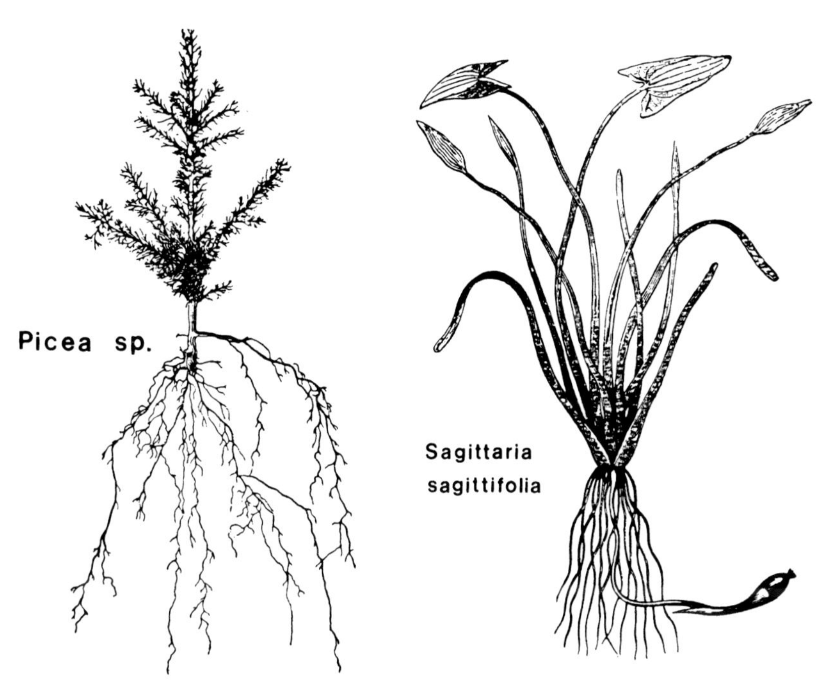

Fig. 14 Branching root system of young spruce tree *Picea* and adventitious roots of arrowhead *Sagittaria sagittifolia*. (From Livingston 1922.)

branches may themselves eventually produce branches, and they likewise, until the whole complex branching pattern of a mature root system is produced.

The overall structure or architecture of a root system is very variable, depending upon soil conditions, such as wetness and fertility, because it results from myriads of responses of individual root tips to the local conditions. For example, branches may be produced more profusely where roots grow through patches of soil which are especially rich in nutrients, perhaps because of a decaying carcass or a fertilizer granule, and these branches grow more rapidly, so producing more branches themselves; in addition, the angle of branching may change. Interestingly, it seems that only plants of more fertile habitats, which might encounter such 'hot- spots' frequently, show this response well; plants from poor soils seem to be less able to respond.

In an annual plant, little more happens to the root system before it dies when the plant itself seeds and dies. An annual plant's root system performs two main functions: gathering water and nutrients from soil, and anchoring the plant, though the latter may be fulfilled as a simple consequence of achieving the former in many cases. There will be some thickening of the older members of the root system, since they must supply and receive supplies from an ever greater number of growing tips and so must have a greater transport capacity. In perennials, however, different considerations apply. Anchorage becomes more of a primary function, especially where large woody trunks develop above ground, and the root system may also become an important store for the plant's food reserves. Many biennials use much of their growth in the first year to produce a massive underground store which is then drawn on for the demands of flowering and fruiting in the following year. As a result, plants such as turnip *Brassica rapa*, sugar beet *Beta vulgaris*, parsnip *Pastinaca sativa* and carrot *Daucus carota* have been utilized over many years as food crops, and cultivated varieties have been selected to have bigger tap- roots.

Perennials, too, store reserves in swollen roots, probably largely because below-ground stores are less vulnerable to predators than those above ground. The solid matrix of the soil has precluded the evolution of a subterranean cow. Nevertheless, these underground stores are liable to be eaten by a large range of invertebrates and smaller vertebrates, and they are frequently protected by toxic compounds produced by the plant as anti-herbivore defences. One of the best known of these is the root of the giant yellow gentian of the European mountains, *Gentiana lutea*, which is used to make a strong-flavoured drink, reputed to have marvellous medicinal qualities. Other familiar, strong-flavoured roots include horse-radish *Armoracia rusticana*, arnica *Arnica montana* and dandelion *Taraxacum*; what is a strong flavour to us and may have pronounced physiological effects, such as the diuretic action of dandelion, is probably lethal and certainly a powerful feeding deterrent to some small animals.

Perennial plants are of two kinds: those that produce new growth each year but die back and pass the winter (or other unfavourable season, such as a dry summer) as some such structure as a bulb, corm, rhizome or rootstock; and those that retain at least some of the previous season's growth and build on it the next year. A bluebell and an oak tree are both perennials and may both survive a very long time, but though a bluebell may spawn many new bulbs vegetatively during its lifetime, none of them ever produces anything bigger than a bluebell; an oak tree gets larger each year. A similar distinction exists below ground. The bluebell produces a new set of adventitious roots each year from the base of the bulb, whereas an oak tree has a perennial root system. Just as the branches of a tree become woody and thicker with age, and bear short-lived leaves, so the roots become thick and woody and bear short-lived rootlets. The main below-ground structure of a tree is just as permanent as what we see above ground, and the parallel even extends to the fact that some woody plants have short-lived rootlets, while on others, particularly evergreens, they survive much longer, often for several years.

Every year, the root system produces a new crop of white, fleshy roots and a proportion of them survive, develop woody tissues and form part of the perennial structure, so that the basic network continues to grow, just as does the branch system above ground. This network acts as a transport system, collecting water and nutrients gathered by the absorbing white roots and moving them to the shoots, and also returning sugars made by the leaves to the roots to provide them with energy and materials for growth. Equally, however, it provides the anchorage for the developing crown above the ground.

The forces acting on a tree can be quite considerable, especially in strong winds, as can be seen by the devastation caused by hurricane force winds such as those of October 1987 and February 1990 in southern England. If the force is too great, the tree may fail either by the trunk snapping or by the

Fig. 15 Sitka spruce tree blown over by a storm in Kielder Forest, Northumberland, showing the shallow root plate. (Photograph G.P.Buckley.)

roots breaking free from the soil. Normally the tree survives because the major woody roots radiating out from the base of the trunk spread the force and dissipate it into the soil. Trees will withstand severe gales only if the distribution of roots and their integration into the soil matrix is sufficiently strong. In some forestry systems, root growth is restricted, especially where land has been thrown into ridge and furrow patterns to aid drainage, for the trees are planted on the ridges and fail to develop roots properly in the furrows. It is now realized that this causes severe windthrow problems in many commercial plantations and foresters are searching for new cultural techniques which will improve root development.

The typical shape of a tree root system is very different from its above-ground crown. Whereas the latter normally has a well-defined trunk, the main roots of most trees radiate out from the base of the trunk; there may also be a few downward growing roots, though they rarely penetrate more than a few metres. When a tree does blow over, this pattern is readily seen: on good, deep soils a pit as much as a metre or two deep may be excavated by the falling tree, the roots pulling the soil up with them. On poorer soils, and especially on waterlogged peat, a fallen tree may scarcely leave any hollow at all: downy birch *Betula pubescens*, one of the commonest trees on peat, has a wholly superficial root system, and Sitka spruce *Picea sitchensis*, which is often grown on peat, is similar and so very susceptible to windthrow (Fig. 15).

Nevertheless, the underground parts of plants can be remarkably exten-

Table 3 Root production in various ecosystems. The data are taken from various sources and are estimates which should be treated with some caution because of the difficulty of extracting roots quantitatively from soil.

Ecosystem	*Root biomass tonnes/ha*	*% of total biomass*	*Root production kg/ha/year*	*% of total production*
Conifer forest	3–85(–209)	15–25	3.5–11	16–73
Deciduous forest	17–95	9–44	5.4–9.0	40
Grassland	16	75–98	5.1	50–85
Cereal	1.2	–	1.1–4.2	49
Soybean	0.6	–	0.6	–

sive, and can make up a considerable fraction of the total mass of a plant. Trees and annual plants (especially crop plants) have the smallest proportion of their mass as roots, with often as little as 10 percent of the total being below ground, although 15-25 percent is more usual for trees and the value may be as much as 40 percent. The actual mass of roots varies enormously between various types of vegetation. For example, a soybean crop may have less than one tonne (1 t = 1000 kg) of roots per hectare (100 m x 100 m), while wheat tends to have over 1 tonne per hectare. Under grasslands there is very much more root, typically 10 to 20 tonnes per hectare, often making up four-fifths or more of total plant mass, and since these roots are generally very fine, the length of root in a given soil volume can be astonishingly high: over 100 centimetres of root per cubic centimetre of soil or around a million kilometres of root under a single field of grass. The mass of roots in forest ecosystems depends very much on soil type, species and (for plantations) the age of the stand, but can be as high as 200 tonnes per hectare. The range of figures that have been reported is shown in Table 3.

These figures suggest that herbaceous plants such as grasses devote much more of their resources to growing roots than do trees and other woody plants, which tend to have rather low below-ground biomasses in relation to what is above ground. It is also possible, however, to estimate the proportion of the annual production that plants commit to above-ground and below-ground growth (Table 3). Whereas grasslands devote resources to roots and shoots in about the same proportion as their biomasses, trees appear to send proportionately much more below ground, since forests rarely have much more than 40 percent of their biomass in roots but always commit at least this much (and sometimes as much as 70%) of production to root growth. There are two reasons for this discrepancy: one is that many fine roots are short-lived and are replaced by new rootlets during the growing season, and the other is that a significant part of the production going below ground in many tree root systems does not actually go to the roots at all, but to fungi which live in asociation with them. These are called mycorrhizal fungi, and they perform many of the normal tasks of the root, especially the absorption of nutrients from soil.

Roots live in company with a very large number of soil organisms, and for many of them the environs of the root is the best part of the soil in which to find food. Roots lose considerable quantities of materials into the soil, both as whole cells sloughed off (for example from the root cap) and as chemicals that diffuse or are secreted from the root. Something like 5-10 percent of all

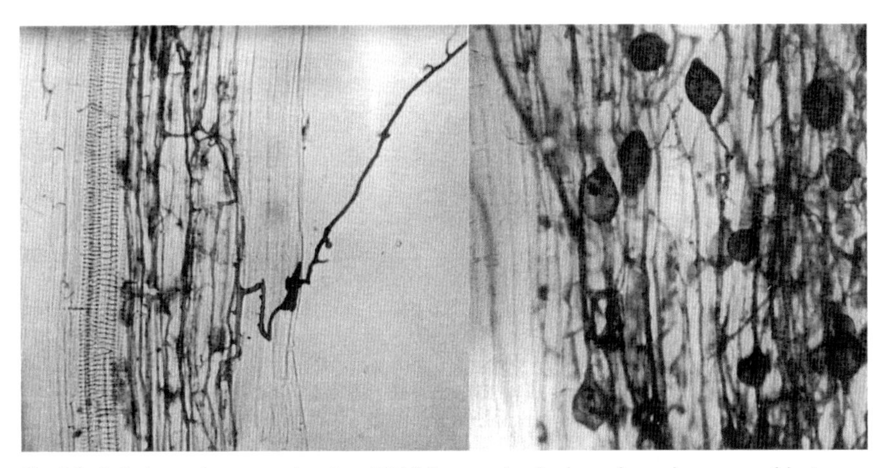

Fig. 16 Left An onion root showing VAM fungus (vesicular-arbuscular mycorrhiza). Entry point with longitudinally running hyphae and arbuscules in the root cortex. *Right* VAM fungus in onion root showing vesicles and hyphae in the root cortex. (From: J.M.Phillips & D.S.Hayman 1970.)

the carbon that is manufactured by the leaves may end up in the soil in this way, even before the root dies. Since the root is the main source of carbon, and hence energy, in the soil, the region of soil around the root resembles an oasis in a desert. This region is called the rhizosphere, and in it develop enhanced populations of bacteria, fungi and some of the smaller soil animals such as nematode worms. Some of the fungi are quite innocuous, feeding on the organic substances lost by the roots; others are pathogenic and may invade the root, damaging it; a third group form symbiotic associations with roots, and these are mycorrhizal fungi.

Mycorrhizas

The word mycorrhiza comes from two Greek words: *mykes* fungus and *rhiza* roots. A mycorrhiza is a root infected with a particular type of generally beneficial fungus, but there are several types of mycorrhiza involving different fungi and plants. The most widespread and most ancient type, though not the most familiar, bears the cumbersome name of vesicular-arbuscular mycorrhiza, almost invariably and understandably abbreviated to VAM. The name derives from two structures formed by the fungus inside the root: vesicles are globular storage bodies, and arbuscules are very finely branched fungal strands (hyphae) where interchange of materials between host and fungus occurs. Both are microscopic, less than a tenth of a millimetre across, and indeed VA mycorrhizas form no visible external structures, so that it is impossible to tell if a plant has the association without microscopic examination of stained roots (Fig. 16). VA mycorrhizas are formed by a small group of fungi, members of the family Endogonaceae, whose principal genera are *Glomus, Gigaspora, Acaulospora* and *Scutellospora*, and these fungi can only survive in association with the roots of a plant. Their principal distinction is the size of their spores (Fig. 17), which are quite enormous by fungal standards – in one species of *Gigaspora* they are over half a

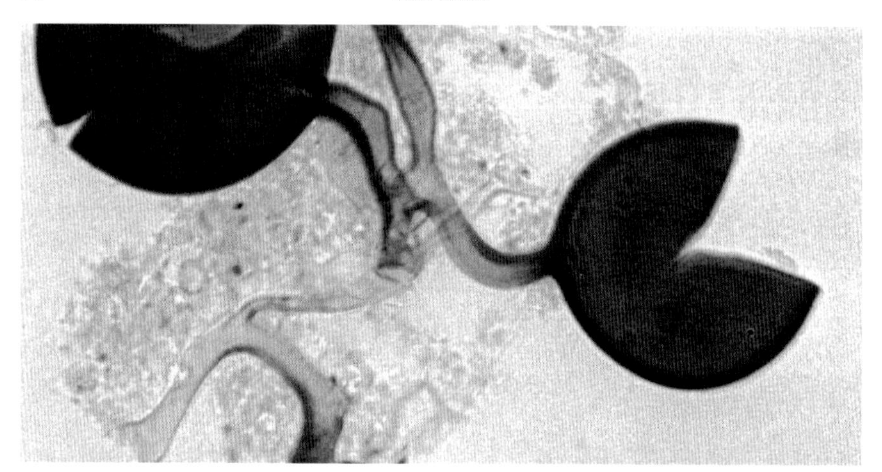

Fig. 17 Spores of the VAM fungus *Glomus invermaium*. (Photograph T.P.McGonigle.)

millimetre across, compared to a typical figure for most fungi of around one-hundredth of a millimetre; in most VAM fungi the spores are about a tenth of a millimetre.

Although the fungi depend absolutely on plant roots, from which they obtain all their carbon for energy, plants can in most cases grow quite well without them. Nevertheless, most plants that can form VAM do so under natural conditions, because the fungus appears to offer a solution to an otherwise severe problem – the acquisition of the essential mineral nutrient, phosphorus. Of the nutrients plants must obtain in large quantities, phosphorus offers the greatest difficulty because it occurs in soil as phosphate ions which are so sparingly soluble that they move only very slowly through soil. Once a root has used up the phosphate in solution in its immediate neighbourhood, it can only obtain more if other phosphate ions diffuse through soil from otherwise unexploited soil nearby. This happens very slowly: a phosphate ion will normally move less than a millimetre through soil in a day. This must in fact have been a problem to the very earliest land plants, since they had extremely poorly developed root systems which would have been much less effective at exploring soil than the highly branched root systems of modern plants. Remarkably, fossils of one of the first land plants *Rhynia*, about 400 million years old, have fungi associated with their rhizomes that appear almost identical to modern VAM fungi; it seems as though the mycorrhizal symbiosis is a very ancient one. Certainly fossils of plants from Triassic rocks (over 200 million years old) have clear VAM fungi in their roots.

Plants appear, therefore, to have faced a severe problem of getting phosphate from soil at a very early stage in their evolution, and they seem to have solved it by associating with fungi on several different occasions. VAM are found on the roots of a wide range of plants, especially herbaceous ones, but also a number of woody plants and trees. Families such as the Ranunculaceae, Rosaceae, Leguminosae, Labiatae, Scrophulariaceae, Compositae, Alliaceae and Gramineae all normally ocur with VAM fungi in their roots. A few families are very rarely mycorrhizal, especially the Cruciferae, Chenopodiaceae and Cyperaceae. About two-thirds of all plants appear to be normally VA mycorrhizal.

There are several other types of mycorrhiza than the VAM, and they clearly evolved quite separately. The best known, and the type which was first recognized, is the ectomycorrhiza or sheathing mycorrhiza, characteristic of many forest trees, especially the Pinaceae (pines, spruces, larches, firs; see Plate 4), the Betulaceae (birches, alders) and the Fagaceae (oaks, beeches). This type of mycorrhiza involves more familiar fungi: almost all are toadstools, members of the Basidiomycetes (see page 122). Some are well known and distinctive, such as fly agaric *Amanita muscaria* which forms a mycorrhiza with birch, and is therefore almost always found growing under birch.

Ectomycorrhizas were the first type to be discovered because they are visibly different from uninfected roots. Ectomycorrhizal roots are stubby and often fork dichotomously, giving dense clusters. Each root tip is surrounded by a sheath of tightly woven fungal hyphae and other hyphae radiate away from this into the soil. It is much easier here than in the VA mycorrhiza to see how the symbiosis works. The fine fungal threads penetrate the soil, picking up the immobile phosphate ions and transporting them back to the sheath, where they are stored before being eventually passed to the root. Meanwhile the fungal hyphae beneath the sheath, which are in contact with the root cells, obtain sugar from them to feed the fungal tissues. It is easy to see what the fungus gains: a reliable source of carbon for which it does not have to compete with other soil fungi. At first sight the gain of phosphate to the plant is clear, too; but why should plants use fungal hyphae to obtain phosphorus for them, rather than growing new roots? The answer lies in the dimensions of roots and hyphae. A root is a complex, multicellular structure, with a central strand of conducting tissue, surrounded by several other layers of cells; the finest roots are around seven hundredths of a millimetre in diameter, but a tenth of a millimetre is more typical. A fungal hypha is only a hundredth of a millimetre in diameter. The cost of building a root or a hypha is proportional to the volume of tissue involved, and this depends on the square of the radius: the hypha is ten times finer than the root and so one hundred times cheaper to build. In simple economic terms, therefore, it pays the plant to harness fungal hyphae to do the job of scavenging for phosphate rather than building roots for the same purpose. The critical point is that it is almost exclusively phosphate that is so hard to obtain: a root system quite adequate for obtaining water, or nitrate or potassium ions, would be incapable of taking up enough phosphate, but if the plant grew more root, the benefits would only be in extra phosphate.

This seems to be the underlying evolutionary rationale for both ectomycorrhizas and VA mycorrhizas, but the former have taken the process one stage further. Fungal hyphae are too fine to carry large volumes of material and they would therefore be unable to supply enough water for a large plant, for example, simply because so much greater a volume of water than of other soil resources is required. The hyphal strands that emanate from ectomycorrhizal roots, however, are remarkably thick, consisting of many hyphae wrapped together. Using electron microscopy (see page 107), one can see that in some cases these hyphae have fused together to create a large diameter channel, perhaps five to ten times larger than an individual hypha. This channel is big enough to move water in large quantities and it seems that some ectomycorrhizal trees do take up at least part of their water needs through mycorrhizal hyphae. This means that much of the resource acquisition function of the root system can be taken over by the fungus, reducing the need for an extensive root system, and this is precisely what occurs in some conifer forests. These seem to have much less root than one would expect; the root system is apparently reduced to the structural skeleton required to anchor the tree and to provide points of origin for the ramifying hyphae, and a limited absorption capacity.

This is not true of all trees, since many of them have VA mycorrhizas, including maples, sycamore, limes, poplars, and elms. Some trees, such as willows, are thought to be able to form either type of mycorrhiza, but the reasons why different trees favour different fungal partners in this way are quite unknown. It does seem that ectomycorrhizal trees are better able to colonize poor soils than VAM trees, and this is probably because the former get more benefit from the more active ectomycorrhizal fungi. Of course, there is a cost to this: the ectomycorrhizal tree probably has to give up more of the carbon it fixes in photosynthesis than does the VAM tree and so the latter may be at an advantage on better soils.

Another remarkable feature of mycorrhizas that has recently come to light is their ability to link plants together. When a root grows it may become infected by mycorrhizal hyphae from older parts of the root or, more commonly, by encountering hyphae in the soil. These latter hyphae are likely to be part of the system radiating from another root, and so the two roots may become linked. Such links may be from root to root in one plant, from plant to plant within one species, or even between species. By labelling trees with radioactive isotopes, it has been found that materials can pass from plant to plant by means of these links. At present it is unclear how much can travel in this way, but there is intriguing evidence that seedlings establish in swards more readily if they become mycorrhizal than if they remain uninfected, and this raises the possibility that they may depend for their survival on nutrients transferred from neighbouring plants. If this turns out to be a widespread and important phenomenon it may force us to rethink our view of plant communities: ecologists have in the past tended to view them as dominated by intense competition between plants; it may be that there is more co-operation than we thought.

There are other sorts of mycorrhizas. Some are obscure and little studied yet, but two are especially worth examining: those on the roots of heathers and of orchids. Heathers form associations with quite a different group of fungi from the ones we have considered so far – the cup fungi, and in particular a fungus called *Hymenoscyphus*. The fungus is found in special fine roots called hair roots which are abundant on the heather root system, and it infects almost all the cells of the root. Although the details of this association are still somewhat obscure, it is clear that it acts in a different way to the main types. The fungus appears to be able to break down organic matter and to transport the nutrients it so obtains back to the host plant. VA and ectomycorrhizal fungi do not seem able to do this. Certainly, the heather mycorrhiza assists its host with nitrogen nutrition as much as with phosphorus, and it is probably no coincidence that heathers are able to grow on extremely nutrient-poor soils through this assistance. The heather mycorrhiza seems to be a special case that has evolved to cope with such extreme conditions, and heathers are more dependent on their symbiotic partner than most plants.

The most remarkable case of dependency, however, is the orchid mycorrhiza. All orchids are mycorrhizal, but here involving basidiomycetes again. Some of these fungi are obscure if abundant (e.g. *Rhizoctonia*), but others are well-known plant pathogens. One of the most remarkable instances is the Japanese orchid *Gastrodia elata* which forms a mycorrhizal association with honey fungus *Armilariella mellea*, which is a virulent pathogen of trees, much dreaded by foresters. Orchid seeds are minute, weighing only a few millionths of a gram, and lack any differentiation into root and shoot. They are virtually incapable of any independent growth and can only successfully germinate if they can associate with their particular mycorrhizal fungus. In their early growth they establish a tuber-like structure called a protocorm, and at this stage they are almost totally dependent on their fungal partner, which supplies them not only with mineral nutrients, but also with carbon for energy, for the orchid has not yet started to photosynthesize. Later, most orchids develop green leaves above ground and the relationship then comes to resemble other types of mycorrhizal association.

Some orchids, however, never produce green shoots, have no chlorophyll and cannot photosynthesize. These non-green orchids are particularly characteristic of very deep shade in forests, where there is insufficient light for green plants to survive. The bird's-nest orchid *Neottia nidus-avis* is the commonest, but others include the ghost orchid *Epipogium aphyllum* (Plate 5) and the coralroot orchid *Corallorhiza trifida*. These are normally described as being saprophytic, feeding on decaying organic matter, which is certainly unusual behaviour for a plant. The truth is even more bizarre. The root systems of these orchids are extremely limited, consisting usually of a short stubby rhizome, or a tangled knot of thick roots (the eponymous 'bird's nest'). They therefore present the unusual situation of a plant effectively without leaves or roots! Other plants like this, such as broomrapes, are parasites, and these orchids are no exception, but whereas broomrapes are parasitic on other plants, the non-green orchids parasitize their mycorrhizal fungi, which supply them with all their needs. This has almost certainly arisen as an extension into adult life of the juvenile dependence of all orchids, but it is certainly remarkable that an immobile plant with virtually no roots and no leaves can parasitize a fungus which is free to grow through soil and may itself in some cases be parasitizing other green plants.

Some other plants can perform the same trick, especially the yellow bird's-nest *Monotropa*. This associates with the same fungi that are mycorrhizal with forest trees, and so parasitizes those trees indirectly. By feeding pine trees with radioactive carbon dioxide, which they fix into sugars, it is possible to trace the path of the carbon first into the pine's mycorrhizal partner, the fungus *Boletus*, and then into the *Monotropa*. Such fungal connections between plants are not unusual, as we have seen already.

Most plants are mycorrhizal, though the types of mycorrhiza and their functions vary widely. As a result there has been great interest in the possibility of using mycorrhizas commercially, but with the exception of commercial forestry, where the trees become infected willy-nilly in most cases, and of specialized horticulture (particularly of orchids), the applications have been disappointing. In most agricultural situations it is cheaper to apply phosphate fertilizer than to inoculate plants with mycorrhizal fungi, and only

Fig. 18 Rhizobium root nodules on broad bean. (Photograph J.Day.)

where the soil has been sterilized to eliminate pathogens, as in citrus orchards in Florida, have the benefits been realized.

Root nodules

Mycorrhizas are not the only association between roots and soil microbes. That between legumes and soil bacteria that can fix nitrogen from the atmosphere has been recognized for much longer and is of enormous agricultural significance. Most members of the family Leguminosae (peas, beans, clovers) have nodules on their roots which contain the bacterium *Rhizobium* (Fig. 18). In the nodules, this can convert the abundant gas nitrogen (four-fifths of the air) into forms that plants, and ultimately all other organisms, can use. As a result such 'nitrogen-fixing' crops have been mainstays of most agricultural systems for thousands of years, and, on a global scale, more nitrogen is still added to farmed systems by symbiotic nitrogen-fixation than by artificial fertilizer.

Rhizobium bacteria are common in soil but they only fix atmospheric nitrogen if they can colonize legume roots. Once inside the root they change form and induce the plant to produce the nodule. There are many strains of *Rhizobium* and some are more effective than others at fixing N, partly because the strains are specific to particular types of legume, so that a strain that fixes N well in bean roots may not do so in clover. The biochemistry of the fixation process is remarkable. Chemists turn N_2 gas into ammonia (NH_3) at high temperatures and pressures. The bacteria have no need of such conditions, for they possess an enzyme, nitrogenase, which can perform the feat at normal temperatures and pressures. One essential, however, is an atmosphere low in oxygen, and this is achieved in the nodule by the plant producing a form of haemoglobin, the pigment found in blood. Just as in blood, the leghaemoglobin (or legume haemoglobin) combines with oxygen, and so protects the enzyme. As a result, effective N-fixing nodules are a distinctive pink colour. The appearance of this pigment in both mammalian blood and legume root nodules is a remarkable instance of evolutionary processes achieving similar solutions to analogous problems.

In many respects the nitrogen-fixing symbiosis works like the mycorrhiza. The bacterium fixes nitrogen from the air and exchanges it for sugars from the plant. Plants benefit from this if the supply of nitrogen from soil is limiting their growth and the loss of sugar to the bacterium is therefore less deleterious. In fertile soils, the extra nitrogen is less valuable and the lost sugar could have been used for more growth, so that legumes lose their advantage and are rare on such soils. Legumes are not the only plants that can fix nitrogen from the air in this way but they are the only plants to use *Rhizobium,* except for one tropical shrub in the elm family. Other nitrogen fixers generally form associations with a different microorganism, an actinomycete called *Frankia*. Whereas the *Rhizobium* association is with a single plant family, the *Frankia* associations have evolved with an apparently random assemblage of plants, including alders *Alnus*, sea buckthorn *Hippophaë*, mountain avens *Dryas* and bog myrtle *Myrica gale*. Altogether, nearly 200 species in 14 quite unrelated genera have evolved this habit. At first, it seems extraordinary that such a beneficial trait should not be more widespread, or alternatively that those plants that have picked up the ability should not be more abundant, since they would appear to have a great advantage over their competitors. In practice, it seems that the benefits are only really great in very nitrogen-poor soils; nitrogen-fixing plants can be abundant on materials that have been deposited by glaciers or on rock debris, for example. One of the first colonists of glacial moraines in many parts of the world is alder, and, in the Canadian Rocky Mountains, the dwarf undershrub *Dryas drummondii*, a relative of the widespread mountain avens *D. octopetala*. Probably this is what happened in northern Europe after the retreat of the last glaciers of the most recent glacial period, some 12,000 years ago. Peat and lake deposits from that period contain a very high proportion of *Dryas* pollen. Similarly, broom and gorse appear abundantly on such nitrogen-poor materials as the sand heaps left behind by china clay mining in Cornwall (see chapter 10).

These plants, however, contain the seeds of their own destruction. As they grow and fix nitrogen from the air with the aid of their microbial partners, so they continually add it to the soil, as their leaves and roots die and decay. In a short time, perhaps 50 years, the level of fixed nitrogen in the soil will have been raised to the point where other plants without this ability can grow quite well. Since these other plants do not have to support a population of bacteria in their roots, they may be able to grow better than the nitrogen-fixing plants and so oust them. Plants with the nitrogen-fixing symbiosis therefore tend to be colonists or plants of other poor soils.

Both legume and non-legume symbioses have as their actual N- fixing partner a bacterium or actinomycete, for in the whole course of evolution, some 3500 million years, only they, with their simple cell structure, have developed the ability to fix atmospheric nitrogen. The whole process of nitrogen cycling in soils is described in chapter 6 but it is worth mentioning here that several other bacteria that live freely in soil are also N- fixers, and some of these are found in the rhizosphere around roots. A tropical grass, *Paspalum notatum*, often has very large populations of the N-fixing bacterium *Azospirillum brasilense* around its roots. When this association was first discovered, it caused great excitement, as it seemed possible that crop plants such as wheat and barley could be bred to harbour these bacteria, so reducing their need for nitrogen fertilizer, with enormous economic and environmental benefits. Unfortunately, recent studies have shown that some initial measurements of N fixation in the rhizosphere were over-estimates, and that there simply is not enough energy (in the form of sugars and other compounds from plants) in the rhizosphere to sustain agriculturally worthwhile rates of fixation. A better solution to the problem of the excessive use of nitrogen fertilizer may be to increase the use of leguminous crops.

Roots in soil exist in association with large numbers of other organisms. Some of these associations are close and symbiotic, such as the nodule bacteria and mycorrhizas, while others are much looser. All of them depend on the fact that the root is the main channel by which energy arrives in the soil; in other words, just as plants are the primary producers that start almost all food chains above the ground, so they are, in the form of roots and also dead leaves, below the ground.

4

The Soil Fauna: Arthropods

A legion of animal life forms inhabit the world of the soil. A summary of this diverse and arcane assemblage, from single-celled protozoa to vertebrates, is given below but a description of all these groups is neither necessary nor appropriate here; this has been given by several authors. Instead, certain groups have been selected whose natural history has been especially illuminated by observation and insight. The following essays make no attempt at uniformity of treatment; indeed quite the reverse: the groups have been selected largely to highlight different aspects of soil biology. To provide some logical structure in treatment, however, this chapter deals with the arthropods – all those animals with an external skeleton and jointed limbs – while the next chapter rounds up with an assortment of other groups ranging from earthworms to moles. The following list is an overview of the soil fauna. The groups in italics are those not found in British soils; groups in bold are those described in the following two chapters:

Protozoa	
Platyhelminthes	flatworms
Rotifera	rotifers or wheel animalcules
Gastrotricha	hairy backs
Nemertini	land nemerteans
Nematoda	**eelworms**
Annelida	**earthworms** (Lumbricidae), potworms (Enchytraeidae), *leeches (Hirudinea)*
Mollusca	**snails** and **slugs**
Arthropoda	
Onychophora	*Peripatus*
Tardigrada	bear animalcules
Crustacea	**woodlice** (Oniscoidea), terrestrial sand-hoppers (Amphipoda, one British species)
Arachnida	**mites** (Acari), **spiders** (Araneae), harvestmen (Opiliones), false scorpions (Pseudoscorpiones), *whip scorpions (Uropygi), true scorpions (Scorpiones)*
Myriapoda	**millipedes** (Diplopoda), **centipedes** (Chilopoda), symphylids (Symphyla), Pauropoda

Insecta	
Apterygota ('primitive insects')	**springtails** (Collembola), bristle-tails (Thysanura), Protura, Diplura (Fig. 3)
Pterygota	about 13 orders of 'higher' **insects** with terrestrial representatives in Britain
Vertebrata	*burrowing snakes* and *amphibians*, **true moles**, *marsupial mole* and *mole-rats*

Soil micro-arthropods

Soil invertebrates span at least four orders of magnitude in size, from the smallest protozoa about 0.02mm long to the largest earthworms which can reach 200mm in this country and considerably more in the tropics. Attempts have been made to classify them into large, medium and small forms – the macro-, meso- and microfauna – rather as soil pores have been divided into macro-, meso- and micropores (chapter 1). It is easy enough to place earthworms in the macrofauna, and soil protozoa, which need a compound microscope to see them properly, in the microfauna. Other groups, however, such as beetles and millipedes, span an uncomfortably wide range of sizes from about 1 to 50mm, and are not so easy to categorize in this way.

The term micro-arthropod, however, is a convenient one to encompass several unrelated groups of small arthropods whose largest members are less than about 5mm in length. They include two major groups, mites and springtails, and a few minor groups which often have no common name, such as the Protura and Pauropoda. The ecology of these latter groups is little known while the mites and springtails have attracted a great deal of attention so we focus here on these two major groups to consider some of their fascinating adaptations to soil life.

Mites

Mites, Acari or Acarina, are the most prolific arthropods in most soils and also the most numerous in species. As their common name suggests, they are very small, usually 0.25-0.5mm in length and rarely more than 2-3mm. Linnaeus, the father of modern classification and nomenclature, may therefore be forgiven, perhaps, for turning a blind eye to their diversity and treating them all as a single genus *Acarus*. A century later, in the 1830s, C.L.Koch published his monumental work on the group. This was not only a major reference source but the terror of subsequent acarologists, for he collected assiduously and drew and named every specimen with scant regard to whether it was just a different sex or an immature form of something already described. We owe a great debt, therefore, to later British pioneers who established the basis of our acarine fauna: to men like A.D.Michael in the 1880s and J.E.Hull, J.N.Halbert and F.A.Turk in the first half of this century.

Mites are related to spiders, harvestmen, false scorpions (Plate 6) and several mainly tropical groups (such as true scorpions) within the class Arachnida. Whereas all these other orders are entirely predatory, mites include many groups which feed on living or dead plant material, bacteria and spores. The diversity of form among free-living soil mites is enormous. Indeed the only obvious features they have in common are their four pairs of legs (in the nymphal and adult stages), and an unsegmented body; and even

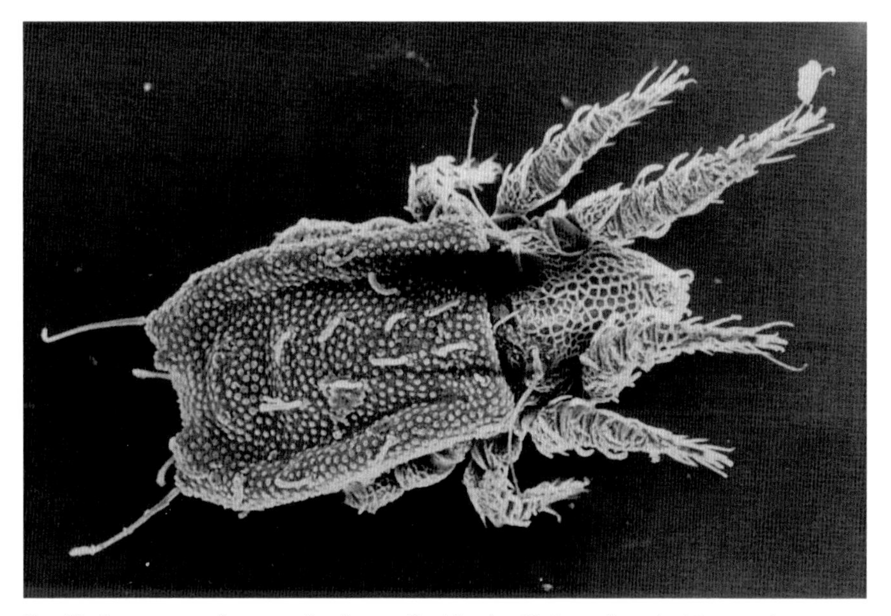

Fig. 19 Stereoscan photograph of an oribatid mite *Nothrus silvestris*. (Cryptostigmata). (National Museum of Wales.)

this may be partially divided in some groups. Eyes are generally lacking but one or two pairs of simple eyes may be present.

The most abundant group of mites in moss, leaf litter and soil are the Cryptostigmata, so called because their breathing pores (stigmata) are inconspicuous. These are known as oribatid or beetle mites in this country and as moss mites (Moosmilben) in Germany. In forest litter they may represent 75 percent of the mite fauna. Nearly all are rather slow moving grazers on bacteria, algae, fungi, spores and dead vegetation. Compared with earthworms, millipedes and woodlice they probably play only a small part in the direct decomposition of litter, but they can be significant in some soils, and more generally important in promoting the distribution of microorganisms which play a vital role in this process.

Most beetle mites are brown and leathery (Fig. 19) or shiny black or reddish (Plate 1) with a tough chitinous skin. Ridges or flaps are often developed over the head with projecting spines which must serve for protection from would-be predators. The oribatid *Ceratoppia* reminds one of the large herbivorous dinosaur *Triceratops* whose horny excrescences presumably played a similar role. Some species carry their cast nymphal skins about on their backs to provide concealment and protection from enemies, or cover themselves with white dust or sticky secretions to which particles of vegetation or dirt adheres. The legs are usually short and stocky and can sometimes be withdrawn under protective flaps. In a few genera such as *Steganacarus* (Fig. 20) the whole front part of the body hinges in the vertical plane and can be folded downwards like a lid against the hind portion of the body after drawing in the legs. When closed in this way, these 'armadillo' mites resemble little, hard-coated seeds or eggs.

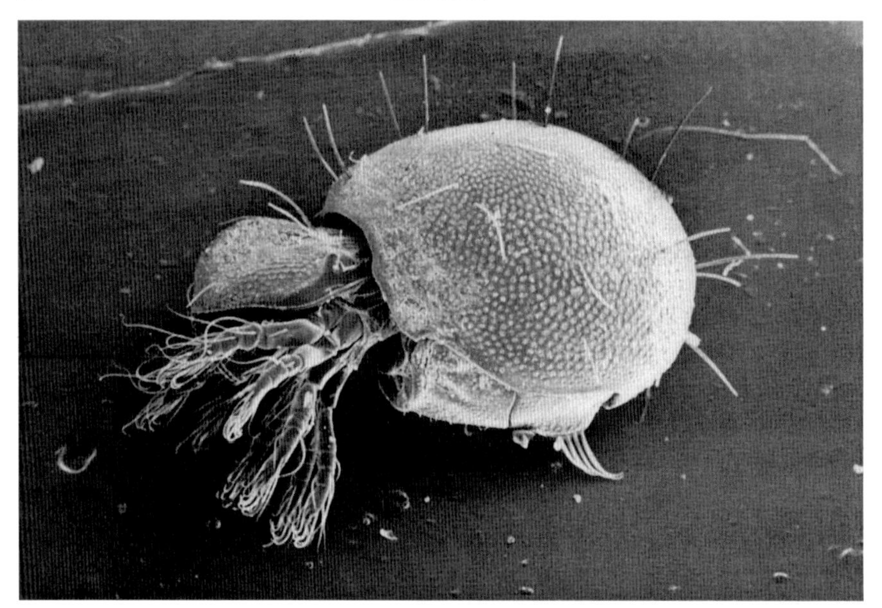

Fig. 20 An 'armadillo' oribatid mite *Steganacarus magnus*. (National Museum of Wales.)

Some very small genera, such as *Suctobelba* and *Minunthozetes*, are very compact and obviously adapted to getting about in the confined spaces of the fermentation layer within leaf litter or within the soil itself. A few species have legs that are longer than the body which makes them look more spider-like (Plate 1). In *Damaeus onustus*, for example, the front two pairs are directed forwards and the hind two pairs backwards so that, although the body is only 1.5mm long, the overall span of the legs is nearly 4mm. This helps it to walk over the rough surface of a woodland floor, carpeted with obstacles.

Oribatid mites shows two types of development very much as in insects. In the 'lower' families, the young stages increasingly resemble the adults, as they do in grasshoppers; in the 'advanced' families the nymphs are quite different from the adult, and there is a marked metamorphosis as occurs in moths, beetles and flies, though without the special pupal stage. A hundred years ago, careful observations showed that the legs of the adult were folded under the body or sides and not formed within the nymphal legs.

The young stages are vulnerable to predation and desiccation, and often live in fungi or plant litter; pine needles, gorse twigs and leaf stalks which have lain on the ground for some time are often riddled with oribatid nymphs (Fig. 21). The duration of life cycles varies greatly, from as little as five weeks to over a year for different species, and there may be one to five generations a year.

The two other major groups of soil mites are the Mesostigmata and Prostigmata in which the breathing pores are situated roughly midway along the sides or near the base of the mouthparts respectively. The Mesostigmata are largely predatory. They include those white or chestnut-brown, fast-running mites that can usually be seen when a sod of grass or handful of garden

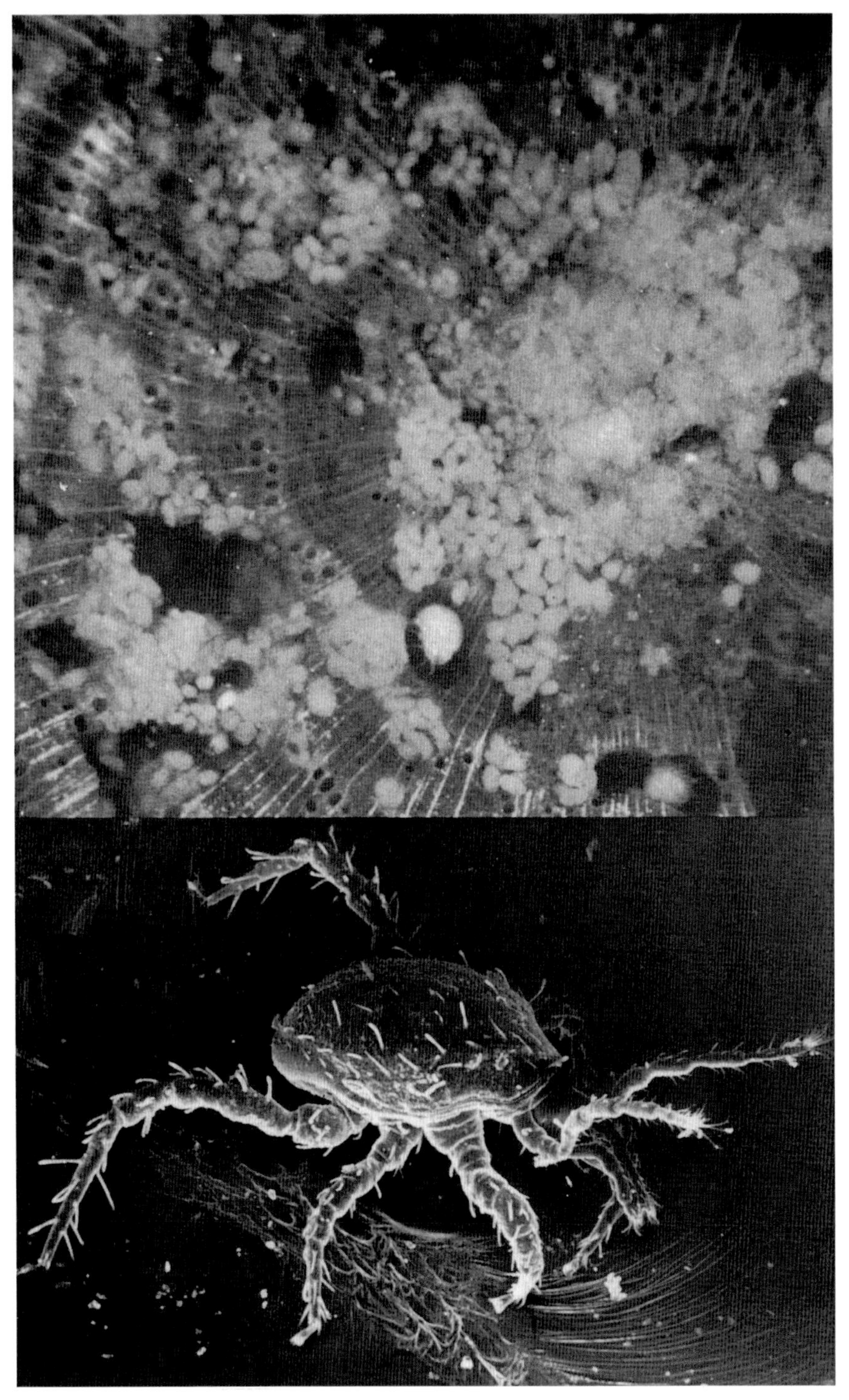

Fig. 21 *Above* Juvenile mites burrowing in a rotten twig in the upper F layer of a beech woodland soil. Note the massses of mite faeces. (J. M. Anderson). *Below* A predatory gamasid mite (Mesostigmata). (National Museum of Wales.)

Fig. 22 A fungal feeding uropodid mite (Mesostigmata). (National Museum of Wales.)

compost is sieved over a tray. The legs are often long and slender, and there is a pair of extrusible mouthparts with scissor-like, toothed jaws at the tip. Their prey consists of springtails, insect eggs, nematodes and other small creatures appropriate to their size. Figure 21 shows a springtail-eye view of such a gamasid mite approaching. Like the oribatids, this group is most abundant in the litter and uppermost soil layers but a few small mites, such as *Rhodacarus* species, are more common a few centimetres down in the soil.

One subdivision, the Uropodina, are fungal and faecal feeders and are commonly associated with ants' nests. They are often circular or oval in outline, with short legs that fold away into grooves under the body (Fig. 22). When not moving, they look like tiny discs from above with little to suggest that they are animate.

The Prostigmata are a heterogeneous assemblage of usually very small and often soft-bodied mites. They are easily damaged by collecting methods that depend on washing soils over a sieve, and often succumb to desiccation before they can escape from samples subjected to heat. The only members of this order that are likely to be seen without a microscope are the plump, scarlet mites, up to 4mm in length, of the family Trombidiidae. They live under stones or in crevices and prey on smaller arthropods.

Collembola

Like the Acarina, the Collembola are one of the mo. ancient groups of arthropods, forming an offshoot from the stock linking millipedes with true insects. They or closely similar forms have been found as fossils in lower Devonian deposits around 400 million years ago, and somc 100 million years before winged insects, such as dragonflies, appeared. Their unique and universal feature is the possession of a ventral tube on the first body segment behind the three pairs of legs. The name of the group, indeed, derives from

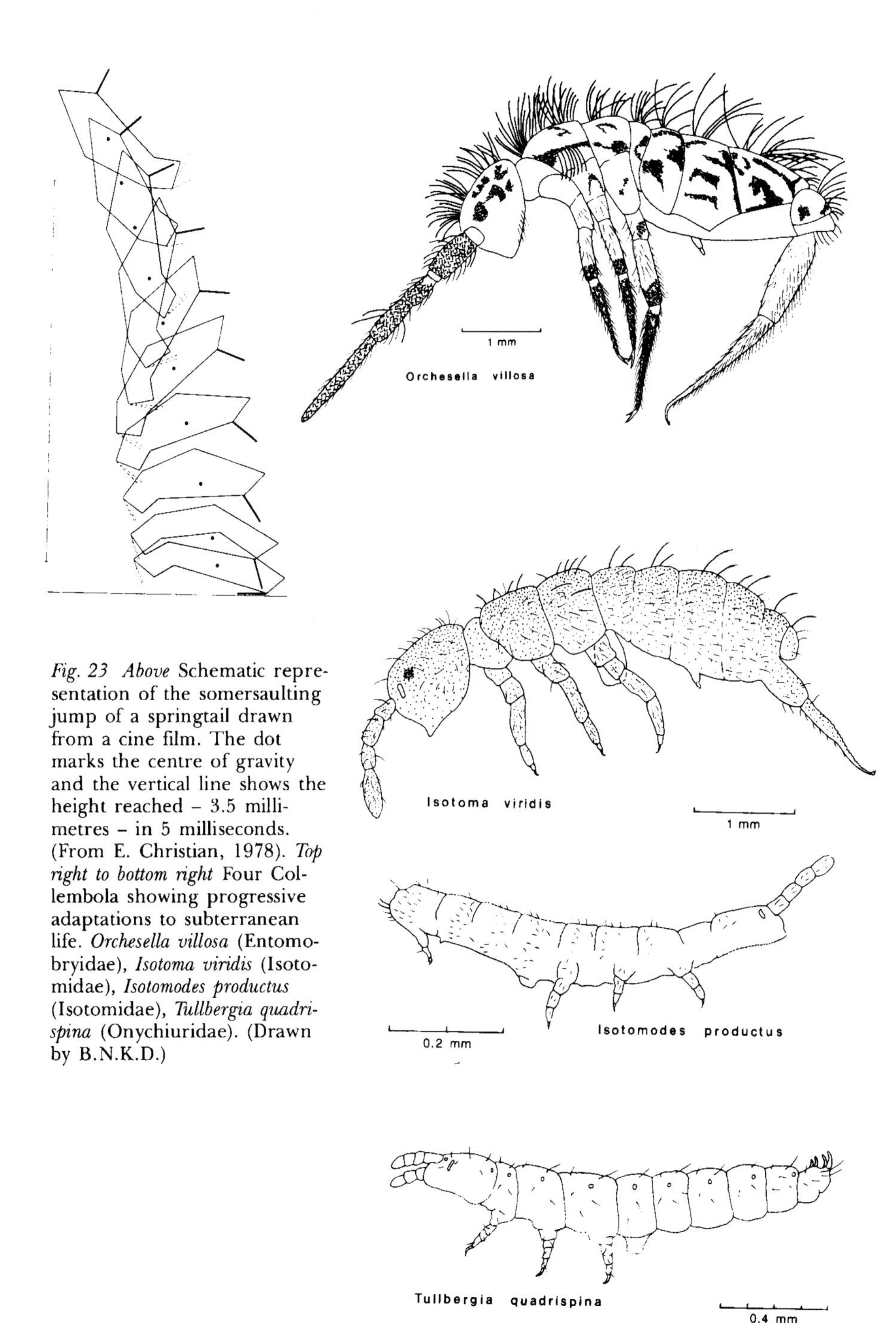

Fig. 23 Above Schematic representation of the somersaulting jump of a springtail drawn from a cine film. The dot marks the centre of gravity and the vertical line shows the height reached – 3.5 millimetres – in 5 milliseconds. (From E. Christian, 1978). *Top right to bottom right* Four Collembola showing progressive adaptations to subterranean life. *Orchesella villosa* (Entomobryidae), *Isotoma viridis* (Isotomidae), *Isotomodes productus* (Isotomidae), *Tullbergia quadrispina* (Onychiuridae). (Drawn by B.N.K.D.)

this feature (from the Greek words for glue-peg) before its function was properly understood. Its main function seems to be to absorb water and ions from the soil or litter into the body fluids.

The Collembola are commonly known as springtails because of another organ, the spring or furca, which is possessed by many species. This arises near the end of the body, and, in its most developed form, is a muscular structure, about the same length as the legs, which is folded forwards under the body and held in place by a sort of retaining clip. When the animal is disturbed, it kicks off and launches itself into the air in a forward somersault, which has been recorded by high-speed cinematography (Fig. 23).

The spring is obviously an important mechanism for escaping from would-be predators, such as mesostigmatid mites – and beetles as described later. However, it is most useful for surface-living species. Even within a single family, the Isotomidae, one can trace a sequence of species showing reduced development of the spring to a small stump or nothing at all. This trend is clearly associated with an increasingly subterranean way of life, and is accompanied by other adaptations such as reduction in lengths of limbs and antennae, reduction in eyes, absence of hairs or protective scales, and lack of pigment (Fig. 23).

Other families show similar adaptations to more or less subterranean life styles, and this has resulted in an ecological classification of Collembola into surface-living and true soil forms. The Entomobryiidae contain many of the large, surface-living species belonging to the genera *Tomoceros, Orchesella* (Fig. 23) and *Entomobrya* itself. Most species of the family Onychiuridae are typical soil forms. They have special glandular cells dotted about the body called pseudocelli (see Fig. 23, bottom) as they were originally thought to be sensitive to light like simple eyes (ocelli). In fact, these appear to be used in defence against predators by producing an unpleasant secretion; many milli pedes similarly produce repellant secretions from paired stink glands on certain body segments.

R.A. Brown gives a graphic description of an encounter between a predatory gamasid mite *Pergamasus lapponicus* and the springtail *Onychiurus armatus*. "When *P. lapponicus* meets an adult *O. armatus* in a void there is little escape for the collembolan. The mite can easily outrun it, and *O. armatus* cannot spring away as it lacks an effective furca. However, when the mite starts to feel *O. armatus* with its front legs, the collembolan stands still and within 2–3 seconds a drop of milky liquid appears on abdominal segment five in the vicinity of a collection of organs known as pseudocelli. The collembolan then bends its abdomen forwards over its head and attempts to smear this liquid around the mouthparts of the attacking mite. If it succeeds, the mite quickly drops the collembolan and backs away, frantically rubbing its mouthparts with its front legs".

Despite such protective and escape mechanisms, predatory mites may normally exert a considerable restraining effect on Collembola populations. A differential susceptibility to DDT is often correlated with an increase in Collembola in sprayed areas (chapter 9).

Collembola show marked seasonal variations in numbers associated with an ability for rapid multiplication when conditions are favourable. Many species feed on fungal mycelia and so thrive in moist conditions in mild temperatures when fungi proliferate. Others feed directly on decomposing plant and animal material. Like many other soil animals, populations often show very patchy distributions in what, to us, may appear quite uniform areas. Collembola aggregations arise partly from clusters of eggs laid in areas with good food resources but scent signals, pheromones, are also secreted which attract passing Collembola into the area.

Woodlice and soil habitats

Woodlice are the only discrete crustacean group to have become widespread in terrestrial habitats. Their closest relatives are all aquatic, and they are structurally less well adapted to life on land than millipedes, spiders or insects. They resemble millipedes in having a calcified exoskeleton which affords protection from enemies, but this is not so well water-proofed; in the more primitive species, up to 80% of water loss from the body takes place through the skin or cuticle. The smallest British species *Trichoniscus pygmaeus* is only 2.5mm in length and lives deep in soil or litter where the humidity is at or near saturation. It loses water so quickly if kept in the open that it only survives a few minutes. The newly hatched young of all species are particularly vulnerable to desiccation and are carried within a brood pouch under the abdomen of the female until they are ready to emerge and fend for themselves. As in their aquatic relatives, breathing takes place through a series of gill-like respiratory plates called pleopods formed by modified abdominal limbs. The genus *Porcellio* and its close relatives have a rudimentary system of branching tubes in the outer pleopods called pleopodal lungs or pseudotracheae, by analogy with the more sophisticated breathing apparatus of insects.

To supplement these physiological features, woodlice have a highly developed behaviour pattern which causes them to seek out places having a suitably high humidity. A few species, including *Cylisticus convexus* and the pillbugs *Armadillidium vulgare* and *A. nasatum*, can tolerate lower humidities and are found in dry habitats such as chalk and limestone scree; *A. vulgare* is, indeed, quite commonly seen wandering about on chalk turf in the sun. Both genera can roll up into a ball; the name of the latter suggests a likeness to the armour-plated armadillos of Argentina, though only the three-banded armadillo can roll up into a perfect sphere (armadillo is a Spanish word meaning, literally, little armoured thing).

Woodlice feed predominantly on dead plant material, and they can play a significant part in the primary breakdown of leaf litter and decaying wood, especially in deciduous woodland. The contribution is mainly in the mechanical comminution of material since as little as ten percent of the food eaten may be assimilated. It seems that digestion is dependent on some preliminary microbial attack having taken place. Without this, a woodlouse surrounded by leaf litter may be in a comparable predicament to the Ancient Mariner with "water, water every where nor any drop to drink". The absorption of copper, which is needed in particularly large amounts, cannot, in fact, be absorbed adequately from fresh food. Woodlice therefore have the unusual habit of eating their own faeces in which bacteria have converted the copper – and possibly other essential salts – into a more digestible form. Rabbits likewise practise coprophagy, and for very similar reasons, while ruminants have perfected the art of double digestion internally.

Deciduous woodland supports the most diverse woodlouse faunas but rough, tussocky grassland can support some of the highest woodlouse

Table 4 Major habitat features from which woodlice were recorded in the non-marine isopod survey. Number of records and % of total.

Habitat	*No*	*%*
Coastal	9943	42
Inland	13535	58
1st ORDER HABITATS	23499	
Garden	2323	10
Lightly/ungrazed grassland	4362	19
Open woodland	2382	10
2nd ORDER HABITATS	13036	
Roadside verge	2493	19
Walls	1650	13
Cliff face	887	7
Shore/strandline	2054	16
MICROSITE	23246	
Stones	9220	40
Litter	2837	12
Dead wood	4300	19
SOIL/LITTER DETAILS	8128	
Oak	552	7
Beech	496	6
Mixed deciduous	1448	18
Grasses	1742	21
Grass/herb	1900	23
HUMUS TYPE	1772	
Mull	1446	82
Mor	326	18

From *Woodlice in Britain and Ireland: Distribution and Habitat*, P.T.Harding & S.L.Sutton 1985.

densities. Populations of 1000 a square metre are common in such habitats, and densities of nearly eight times this figure have been recorded for *Trichoniscus pusillus*. This, the commonest British species, occurs throughout the British Isles including the Highlands of Scotland. Its exceptional ecological versatility may be aided by the existence of two races or forms which are distinguishable genetically: in addition to the normal, diploid form *T. pusillus provisorius*, there is a parthenogenetic, triploid form (with 3 instead of 2 sets of chromosomes) *T. pusillus pusillus*. The two are very alike in appearance so populations are best distinguished by the virtual absence of males in the *pusillus* form. They cannot, of course, interbreed. It is interesting to speculate on the evolutionary advantage of triploidy and the abandonment of sexual reproduction. The *pusillus* form appears to be the more common and widespread of the two, and it was thought formerly that it might altogether replace the *provisorius* form in the north of the country. However, *provisorius* is predominant in some sites in Yorkshire and North Ireland, so it is now thought that warm, well drained habitats, such as south-facing rock scree, may favour this form.

The distribution of these two forms was worked out during the Isopoda Survey Scheme, launched in 1968, through the use of a recording card listing most of the terrestrial and freshwater species and a selection of marine

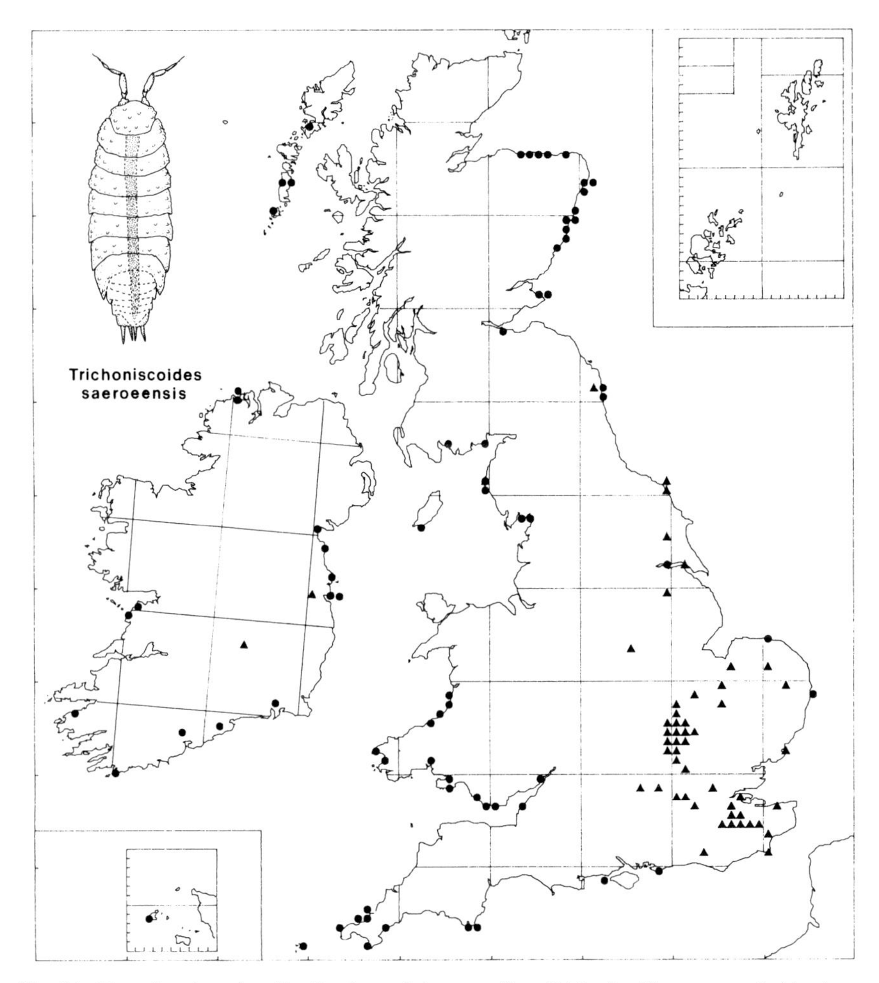

Fig. 24 Map showing the distribution of the woodlice *Trichoniscoides saeroeensis* (dots) and *T. albidus* (triangles). (Data compiled by Isopoda Study Group, drawing by S. V. Green.)

species. There were already many other similar recording schemes in being, backed by enthusiastic volunteers and coordinated by the Biological Records Centre at Monks Wood. But this scheme pioneered an ecological approach. Instead of simply mapping the distribution of species on a 10 x 10 km square grid, the isopod record cards listed several major and minor habitats and habitat characteristics which would provide insight into the ecology of the species.

A little later, this approach was adopted for millipedes, centipedes and woodlice in a common habitat classification which specified a series of first and second order habitats, microsites, habitat qualifiers and soil/litter details. A selection of the 167 boxes which were most often ticked illustrates the features that woodlice most often seek (Table 4). The scheme provided a great stimulus to recording so that, by 1982, a total of 23,499 woodlouse species records that included some habitat data had been analysed.

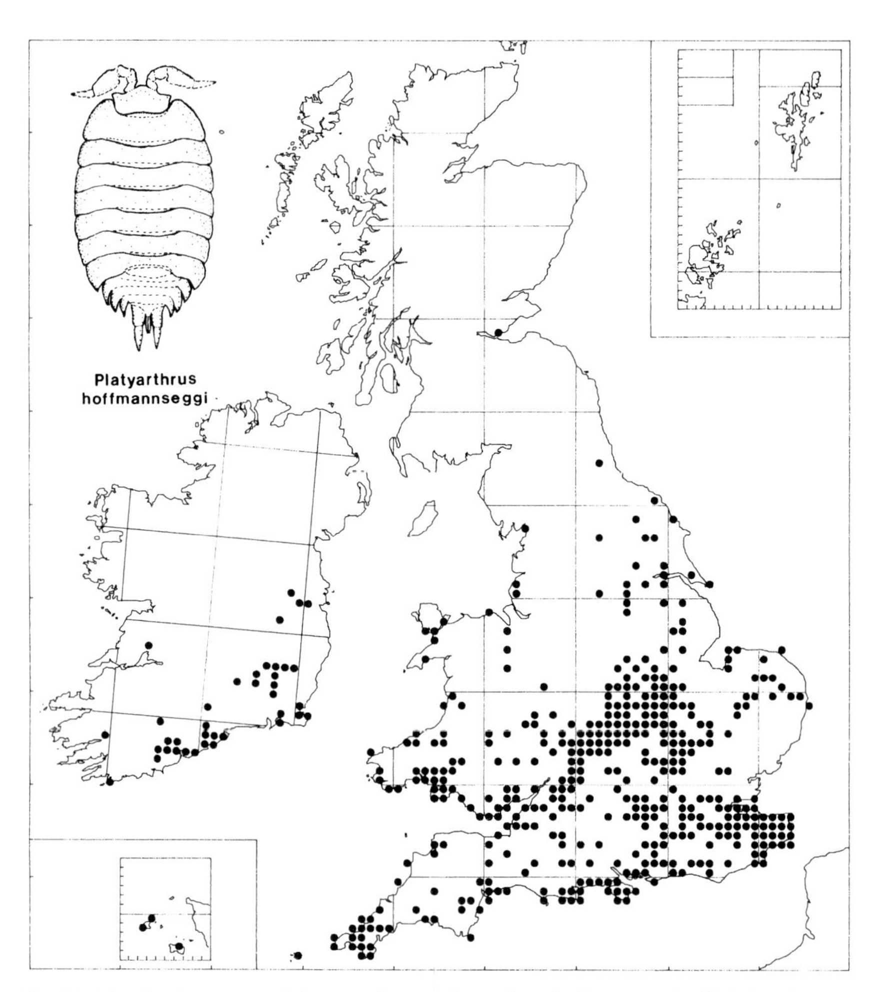

Fig. 25 Distribution map of the woodlouse *Platyarthrus hoffmannseggi* which is often associated with nests of the yellow ant *Lasius flavus*. (Data compiled by Isopoda Study Group; drawing by S.V.Green.)

This compilation resulted in a book which gave a distribution map of each species in the British Isles and quantitative data on the habitats in which it had been found. Maps showing the distribution of three soil species are shown here. There is inevitably a good deal of bias in the records as some places are searched more easily, such as litter, dead wood and stones, as compared with places like rock crevices, soil or shingle. Nevertheless, one can now pick out about 18 species that are often found at some depth in the soil.

Seven of these species are entirely or predominantly coastal. Together with the intertidal sea-slater *Ligia oceanica*, which is widespread on rocky coasts, they suggest a way in which the group as a whole may have colonized the land from fully marine habitats to the supra-littoral zone. These coastal species are found on sandy shores, shingle, boulder beaches, salt marshes or sand dunes. Sand dunes form a bridge to inland habitats, for 20 species are found here, nearly 60 per cent of the British fauna.

Buddelundiella cataractae was first discovered in Britain in 1983 in a garden in Cardiff, but has since been found in shingle, down to 80 cm deep, on the north west coast of Norfolk. This suggests that it may occur elsewhere around the coast. Being white, however, like many soil species, and rarely above 3 mm in length, it is difficult to see among sand grains, especially when it rolls into a ball. Moreover, shingle is a particularly difficult material to search, so without some clues to likely microsites the prospect of finding such a rare species is daunting. *Trichoniscoides saeroeensis* has been found more widely around the coast of Britain and Ireland (Fig. 24). It is essentially a soil or deep litter woodlouse of the supralittoral zone at and a litle above extreme high water mark. This species is often quite abundant once you find the exact microsite.

What appeared to be an obvious case of recorder bias was seen in the early records of *Trichoniscoides albidus* (Fig. 24). This species was widely recorded in Bedfordshire but rarely elsewhere. The distribution pattern reflected the activity of one individual, A.J. Rundle, and his particular ability to recognize and search suitable sites such as wet soil on the sides of ditches and culverts and on stream and river banks. Similar collecting methods applied in other counties by other competent recorders produced only a thin scatter of records. More recent records in the London area and in Kent are still largely due to his efforts but there are now other records from East Anglia and up the east coast.

The last map (Fig. 25) shows a southern or lowland bias which is seen to a greater or lesser extent in several species. *Platyarthrus hoffmannseggi* is an interesting case as 54 per cent of its records are from ant nests, especially of the yellow ant *Lasius flavus* whose influence on soil is described later. The nature of the association is uncertain, but this little woodlouse may feed on ant faeces and so help to clean the nests.

Millipedes and centipedes

Most people can distinguish common millipedes, centipedes and woodlice by their general body proportions and movement. It is surprising, therefore, to find that the 1981 Oxford Pocket Dictionary defines millepede *sic* as a "small, many-legged crawling animal: wood-louse." Zoologists place millipedes and centipedes in the Myriapoda while woodlice are grouped with shrimps and crabs in the Crustacea, so separating the two classes as distinctly as birds, mammals and reptiles among vertebrates. In passing, we should mention two other relatively unfamiliar groups of small myriapods, the Pauropoda and Symphyla, though the glasshouse symphylid *Scutigerella immaculata* (Fig. 26) is a well-known pest and is notable for its habit of penetrating to a depth of two metres or more into the soil.

Millipedes

There are 53 species of British millipedes of which five have been added in the past fifteen years. They are predominantly animals of the forest floor and feed on plant detritus including dead wood. Up to a dozen species have been found in oak woodland on limestone soils with densities of 350-650 a square metre, rising to 850 when broods of young first appear. Fewer species inhabit

Fig. 26 Glasshouse symphilid *Scutigerella immaculata*. (Photograph Broom's Barn Experimental Station.)

base-deficient soils but here, in the absence of earthworms, they sometimes play an important role in the breakdown of plant matter, and produce characteristic moder humus through the accumulation of their faecal pellets. Several species inhabit rough grassland or heathland, and only a few true soil species occur in arable land. Manured soils encourage the build up of populations of the spotted snake millipede *Blaniulus gutulatus* which can reach 1600 per square metre locally and do considerable damage to root crops.

The vernacular name millipede implies a thousand feet but the maximum for species in this country is a little over a hundred pairs, and the short and broad pill millipede *Glomeris marginata* only has 19 pairs. The technical name Diplopoda refers to the fact that most segments appear to have two pairs of legs. This arises from the fusion of pairs of segments as an adaptation to a burrowing way of life. These diplo-segments are linked by elaborate ball-and-socket joints which allow for flexibility of the body as a whole while preventing the telescoping of segments during the low-geared thrusting power of the numerous legs. The legs spend more time on the ground propelling than off the ground, and the ability to push is characteristic of the Diplopoda.

The most familiar millipedes are the cylindrical black species up to five centimetres long with 30-60 rings. When moving freely on the surface, their short legs move in a series of six or seven waves down the length of the body. At rest, they commonly roll into a plane or helical spiral, depending on the species, and display a fringe of projecting legs. Many species also curl up when disturbed but some freeze like players in musical statues while the snake millipede *Ophiulus pilosus* thrashes violently from side to side. These millipedes of the families Julidae and Blaniulidae comprise about half the British species, and a much higher proportion of individuals found actually in the soil.

Four species, including the common pill millipede *Glomeris marginata* (Fig. 27), are able to roll up into a perfect ball with all their limbs concealed. In this state *G. marginata* may easily be mistaken for the pill woodlouse *Armadillidium vulgare* mentioned earlier. It is common in calcareous districts both in grassland and woodland. Feeding studies with this species indicate that only 6-7 per cent of the ingested material is assimilated; its main contribution to the

Fig. 27 Pill millipede *Glomeris marginata* (Photograph J. Grant.)

soil economy is, therefore, in the physical comminution of leaf litter and dead grass which is thus rendered more susceptible to microbial attack.

The flat-backed millipedes in the family Polydesmidae form another major group (Fig. 28). They are characterized by horizontal keels which project on either side of each segment and which act as wedges in splitting apart compacted leaves. One of the smallest British millipedes belongs to this family while rejoicing in one of the longest names, *Macrosternodesmus palicola*. Fully grown individuals do not exceed 4 x 0.4mm in size.

Millipedes have a life span extending over several years. Eggs hatch into six-legged larvae which resemble insect larvae in many ways and suggest a common origin. The number of segments and legs increases with each moult until maturity is reached. In some species the number of eyes also increases in a distinctive pattern which allows one to determine the particular stage, or instar, and to distinguish otherwise very similar species. Two distinct breeding strategies are found among the Julidae: adult *Julus scandinavius* and *Ophiulus pilosus* breed only once and then die, whereas the "black, swift-footed millipede" *Tachypodoiulus niger* and the handsome orange-striped *Ommatoiulus sabulosus* breed and moult several times. The males of these latter two species show a curious alternation of reproductive and non-reproductive stages termed periodomorphosis. This is thought to be an adaptation which allows the dispersal of both mature males and females. Whereas all mature females are capable of breeding, a male which has bred cannot normally moult directly into another breeding male, presumably because of the physical difficulties of re-forming the modified legs, or gonopods, that transmit the sperm. The 'intercalary' stage opens the way to repeated reproduction by males without which there would soon be a great imbalance of sexes

Fig. 28 'Flat-backed' millipede *Polydesmus angustatus*. (Photograph J. Grant.)

among mature individuals. A few species avoid this particular problem by reproducing parthenogenetically; males are either rare or absent altogether.

Estimating the numbers of such creatures is easy enough, if laborious, when they can be washed out of a friable soil over sieves and floated off on a solution of magnesium sulphate which has a high density (like the Dead Sea). With soils containing a lot of organic matter, this method is not possible, and it is usual to use a heat and humidity gradient to expel them from soil cores. However, this relies on a thorough understanding of behavioural responses to such factors, and experiments show that there are often subtle changes in activity even between different juvenile stages of the same species. Pitfall traps are good at catching the adults of many litter-inhabiting millipedes but the young stages of *Polydesmus* species, for instance, are rarely caught this way.

Garden compost heaps and piles of leaves in city parks make good hunting grounds for millipedes, woodlice, pseudoscorpions and other litter-loving invertebrates. Some exotic species have been added to the British list in recent years from such habitats, including *Unciger foetidus* and the blind *Cylindroiulus vulnerarius*. They have probably been introduced from the Continent, or even further afield, on the roots of imported plants. They are said to be synanthropic since they are associated with human habitation and do not seem to colonize rural habitats. Perhaps they can simply cope better than the native fauna with the exotic diet produced by imported trees and shrubs.

Centipedes

Centipedes may be thought of as the predatory counterparts of millipedes. They feed mainly on small insects, mites, spiders, nematodes and other centipedes. Some vegetable matter is consumed and one species *Haplophilus subterraneus* is known to turn vegetarian at times, and cause damage to root crops. The most characteristic and unique feature of the group is the pair of large, inward curving poison claws borne on the modified first pair of limbs under the head. These can clearly be seen with the naked eye when looked at from below. *Lithobius pilicornis* can inflict a painful bite to humans, while the large, tropical *Scolopendra* centipedes, which frequently measure 15 cm in length and more than a centimetre across the body, can partly immobilize small mammals with their poison.

Apart from a few small relatives of this tropical genus, there are two main groups of centipedes in Britain: those living mainly on the soil surface and those living underground. The somewhat forbidding names Lithiobiomorpha and Geophilomorpha merely mean stone-living forms and soil-loving forms. The lithobiid centipedes are relatively broad animals with 15 pairs of long, powerful legs adapted for fast moving. Large specimens may reach 30-35mm in length and span 15mm across the extended legs. Although all centipedes lack the calcareous and waxy exoskeleton of millipedes, and are therefore liable to desiccation, they undoubtedly roam widely at night in search of prey while seeking refuge under stones and in litter during the day.

The Geophilomorpha are less often noticed but are the largest group with 23 species (plus two or three more currently being described). They are recognized by their very elongated serpentine body and curious flickering movement of the antennae when in motion (Plate 7). Most species are between 20 and 60mm long when adult, and bear 37 to 101 pairs of short legs – much as in millipedes. They are red, brown or pale yellow in colour, and some species are phosphorescent particularly in the autumn and when disturbed. They have no eyes but are able to perceive light through their thin leathery skin.

Their most striking feature is their enormous flexibility which is achieved by subdivision of each segment into two – the opposite anatomical condition to that seen in millipedes. With the extra joints, an animal can turn back on itself by the flexing of three adjacent segments, a facility which allows it to explore narrow crevices and interstices in the soil. For reversing out of blind alleys that are too narrow even for this, the rearmost pair of legs are adapted to feel the way like a pair of posterior antennae. An ability to telescope the segments into one another also enables an animal to elongate and contract quite markedly with corresponding constriction or expansion of the body. This helps it to burrow through loose soil in much the same way as an earthworm.

The development of lithobiids is similar to that of millipedes in that the number of legs increases with every larval moult. In the other two groups, the adult number of legs is present from the first stage onwards. Truly, Nature has tried most alternatives. After hatching, the tiny geophilids are unable to fend for themselves, as young lithobiids can, and they depend on their yolk-mass for the first six to eight weeks. During this period, they remain in a compact group in the brood chamber excavated by the female who wraps herself round them thus maintaining a high humidity and affording protection. The total life span is five or six years, a surprisingly long time for such apparently delicate creatures.

Spiders

There are some 632 species of spiders in Britain, divided among 31 families, but few of them penetrate the soil to any extent and so they have generally received perfunctory treatment in soil biology texts. It is true that most spiders live above the ground altogether, weaving their webs in low herbage and bushes, or climbing on trees or walls. Nevertheless, many are to be found living among leaf litter, in cracks in the ground and in the larger spaces around grass roots, under stones and even on bare ground. Here they serve as an invisible brake upon the populations of springtails, flies and other small arthropods that live in these superficial habitats. We are largely ignorant of their importance in the soil economy, as indeed we are about most predators, but a start has been made in recent years in determining their contribution to the energy dynamics of ground-living communities.

There have been several attempts to estimate the size of spider populations of various habitats based on straightforward searching within a square frame placed on the ground, by extracting from soil cores using washing and sieving techniques or heat, and by pitfall trapping. None is entirely satisfactory because the more active kinds of spiders tend to be overestimated in pitfall traps and underestimated in soil and litter samples. More important, though, is the structure of the litter layer and the time of year at which the samples are taken. A grass ley grazed and trampled by cattle has a very poor spider fauna. On the other hand, an undisturbed grassland or fen with a good layer of decomposing litter can support a very high population. Gordon Salt and colleagues at Rothamsted obtained an estimate of up to 130 a square metre from an area of pasture land after first cutting off the vegetation close to the surface. By comparison, Eric Duffey obtained estimates which ranged from 800 to 1000+ a square metre in September and October from an area of limestone grassland with a well developed litter layer at Wytham Wood, Oxford. Spread out evenly on a grid, this last density would be equivalent to about one spider every three centimetres but in the three-dimensional turf environment it would not appear so great.

Spiders may be divided into four groups, or 'guilds' to use the ecological term, according to the way in which they catch their prey: web spinners, jumping spiders, ambush spiders, and wandering or running spiders. In the first group we have the money spiders, Linyphiidae, which feature predominantly in many samples from arable soil, turf and litter. This large group of small spiders abounds from the sea's edge to the tops of our highest mountains. It includes some of nature's most bizarre creations with stalked or lobed 'heads' bearing a ring of eight eyes for all-round vision, as befits hunters which are also prey to many larger predators in the dense micro-undergrowth of grass turf. Linyphiids are responsible for the hammock-like webs and carpet of gossamer that sometimes adorns the grass on dewey mornings. Ground-living species may only spin rudimentary webs or mere tripwires across soil crevices or in leaf litter as described in chapter 1. The young of several abundant species have perfected the art of parascending on silk threads as a means of dispersal. Such aeronautic activity takes place all through the year but is most noticeable in autumn when the population densities are highest.

The significance of this behaviour is illuminated by some interesting observations on the spider fauna of sewage filter beds by E. Duffey. The beds provide an artificial subterranean environment which supports high densities of some linyphiids. *Leptorhoptrum robustum* was recorded at the Birmingham Sewage Works at a mean density of 29,000 per cubic metre over a 28 month period. The deep layer of clinker affords numerous small air spaces where temperature and humidity remain fairly constant, and where sewage input provides a rich nutrient resource. These conditions support a high population of enchytraeid worms and fly larvae on which the spiders feed. When the flow of sewage is shut off for maintenance work, the bed begins to dry out. This results in massive mortality of the prey species, and triggers a wholesale emigration of the spiders. The surface of the filter beds becomes covered with silk at this time, and workmen have reported bites from the spiders which cause irritation and swelling. *L. robustum* occurs widely in marshy areas in Britain where periodic drying out would create the need for dispersal to more favourable sites.

There are several ground-living spiders that use webs to line burrows under stones or down large cracks. Here they wait to pounce out upon passing insects that make contact with the silk threads on the surface. The purse spider *Atypus* constructs a closed finger-like tube which lies flaccid over the soil surface, and may extend for 20 centimetres from the burrow. When a beetle, earwig or other insect walks over the web, it is stabbed from below by the enormous vertical fangs of *Atypus* which then cuts a hole in the tube and drags the victim inside. Several tropical relatives of this mygalomorph spider build trap-doors across their burrows into which unsuspecting prey can be precipitated.

The spiders which jump to catch their prey all belong to the Salticidae. They are diurnal predators with keen eyesight. Most species are active climbers on tree trunks, fences and walls, and are therefore outside our scope. However, a few are common in woodland and fen litter or on stony ground. The crab spiders, Thomisidae, represent the 'ambush guild'. They are characterized by the lateral extension of their legs and their ability to scuttle sideways as well as forwards and backwards. Species of *Xysticus* and *Oxyptila* are often found lurking among fallen leaves, waiting to pounce on any small insect that comes within their reach.

The wandering or cursorial spiders belong chiefly to four families, but a few from other families have abandoned the web-building habit; *Pachygnatha*, for example, is a representative of the familiar orb-web spiders which has adopted a free-ranging life style. Many are nocturnal hunters that shelter by day in silk-lined cells under logs or stones. These range from the tiny *Oonops pulcher*, whose name means beautiful egg-eyes, and which is only 1.5-2.0mm long, to the fearsome *Dysdera* and *Drassodes* which may reach 15mm or more in length. *Dysdera* is unusual in its penchant for woodlice; it is able to pierce the heavy armour of *Porcellio* and *Oniscus* as well as that of the more generally palatable *Philoscia* (Plate 8).

The largest group of hunting spiders are the wolf spiders, Lycosidae. The females are often noticed in summer as they scramble over rough ground and through vegetation with a large pale egg sac held securely in their spinnerets at the end of the abdomen. Without this, the dull grey or brown colour of their bodies renders them very inconspicuous on bare ground or on a woodland floor. The family includes thirteen species of *Pardosa*, all about

5 or 6mm in length and adapted for sprinting to catch their prey. This genus provides a good example of the way in which morphologically similar animals 'share out' the environment in a way that reduces competition between them through a combination of geographical and habitat preferences. Some species are restricted, presumably by climatic features, to northern or southern counties in Britain, while a more finely-tuned discrimination is achieved by staking a claim to woodland, mountain, fen or coast, and the selection, within these, of damp versus well-drained soils or exposed versus sheltered sites. "Every species" says Paul Collinvaux "has its niche, its place in the grand scheme of things......Wolf spidering is a complex job, not something to be undertaken by an amateur".

Vegetarian insects

Many groups of insects live on or in the soil for at least part of their active lives. Even if the adults are free-flying creatures which exceed our earth-bound view, their larvae may play an important role in the soil economy – and sometimes in man's economy too. This section concentrates on those with a mainly vegetarian way of life: aphids, crickets and their allies, and the larvae of some moths, flies and beetles. Passing mention should be made in this context to some booklice and larval thrips (thunderbugs) that live in the litter. These last two groups lead generally unobtrusive lives in comparison with most of the other groups mentioned in that they do not force themselves on man's attention by attacking crops. Termites are a large and important group in the tropics, both for their influence on soils and for their damage to wooden structures, but they do not occur naturally outside the Mediterranean area in Europe.

Greenfly or blackfly are familiar to most people as plant lice which often form dense colonies on the stems and leaves of plants. It is not so well known that several species of aphids live underground on the roots of plants tapping the sap in the same way as their above-ground relatives. Some species are quite specific in the host plants they attack, such as *Aphis cliftonensis* on the roots of rock-rose. Others attack a range of related or even unrelated plants. The tulip-bulb aphid, for instance, occurs on the bulbs or corms of tulip, iris, gladiolus and crocus. Several species have a curious alternation of generations, one of which is spent on a tree and the other on the roots of one or more herbaceous plants. Thus the pear-hogweed aphid and the poplar-lettuce-root aphid divide their attentions as their names indicate; the wingless generation of the dogwood aphid lives on the roots of grasses and cereals, and the elder aphid *Aphis sambuci* migrates in summer to the roots of docks and various pinks and campions in the Caryophyllaceae.

Many of these root aphids are attended by ants. Thirteen species of subterranean myrmecophilous aphids were identified in a study on Staines Moor in Surrey. They were all associated with the common yellow ant whose ant-hills are described later. The aphids were clustered in cavities made by the ants, and it was estimated that the ants consumed over 3,000 of the youngest stages a day in the summer, as well as feeding on the carbohydrate-rich honey dew excreted by the aphids. This phenomenal cull was made possible by the ability of the adults to produce four or more young a day. The aphids may, indeed, have provided enough food to sustain the ant populations with little need for other prey. In return the ants protect the aphids from other predators that would be less concerned with sustainable yields.

Frog-hoppers and their allies are closely related to aphids, and some of these have subterranean larvae which feed on plant roots. The boldly coloured black-and-scarlet adult *Cercopis vulnerata* is commonly seen on nettles and other vegetation in the early summer. A very much rarer insect is our one species of cicada *Cicadetta montana*. This is confined to certain areas of the New Forest in Britain though it ranges widely through Europe, southern Scandinavia and Asia, probably to the Pacific coast. Cicadas are famous as adults for their musical 'singing' which is analagous to the stridulation of crickets. They are also famous for their long larval life. An American species *Magicicada septendecim* is known to live for 17 years as a larva before emerging from the ground in synchronized adult swarms. It is not known for certain how long the British species usually lives its secretive subterranean life. J.A.Grant made a long-term study of this species in the New Forest. Here, 6 and 7 year cycles were observed and the full range of variability is probably 5-8 years. The larvae (Plate 9) feed by sucking the root sap of various plants, most notably purple moor-grass, but also bracken, ling, beech and birch. In parts of the USSR, there are reported densities of 10,000 larvae per cubic metre of soil, and forest trees are severely damaged, root and branch! In Britain, it is a protected species.

The mole cricket is a handsome insect 35-40mm in length when full grown (Plate 10). It is notable for its remarkably broadened and toothed fore limbs which recall the specially modified 'hands' of the mole. Its latin name *Gryllotalpa* is a combination of the latin names for cricket *Gryllus* and the European mole *Talpa*. It lives mainly in underground burrows in grassy swamps and undrained pastures where it feeds on roots and also on insect grubs. Like the cicada and our three species of native litter-living cockroaches, it is on the edge of its range in this country, and is vulnerable to a recession in the climate. It is now restricted to parts of Wiltshire, Hampshire, the Isle of Wight and east Sussex, and, with the cicada, has recently been placed on the list of endangered British species. Land drainage and pasture improvement have reduced the areas of suitable habitat so that, nowadays, there is less scope for recovery after local extinctions. It is interesting to contemplate the possible increase in such species as a result of climatic warming through the 'greenhouse effect'. Ecologists are now seriously considering the possible implications of a 4–5°C rise in average temperature, unprecedented in the historical period, by the year 2050.

There are only a few moths whose larvae live in the soil. The tiny *Micropteryx* species are considered to be very primitive because they retain mandibles in the adults. The larvae are only a millimetre or two long and feed on fine particles of litter in woodland soil. The swift moths are also primitive. The elongate white caterpillars of the ghost swift *Hepialus humulis* attain a length of nearly five centimetres and burrow deeply in the soil feeding on the roots of grasses and a wide range of weeds. They, and the smaller garden swift caterpillars, are very active when disinterred, and are able to crawl backwards as quickly as forwards. The ghost swift is widespread throughout Britain with populations of caterpillars in agricultural land often between 2,000 and 4,000, and sometimes up to 20,000 a hectare. Locally they can cause considerable damage to cereal crops, root vegetables, strawberries and hops, but they are readily controlled with insecticides these days.

The other main pest species are the night-feeding noctuid larvae, known as cutworms from their habit of biting through the stems of plants at or just below soil level. They include the turnip moth *Agrotis segetum*, garden dart *Euoxa nigricans* and large yellow underwing *Noctua pronuba*, all of them voracious and impartial feeders on young plants. They can do a lot of damage to seedlings of lettuce, beet, carrots and onions and even young trees in nurseries.

About a quarter of the British insect fauna belongs to the order Diptera which includes flies and midges – as many as the beetles and moths put together. There are at least thirty families whose larvae commonly occur in the soil, and they are often quite abundant. In general, however, they have tended to be the Cinderellas of the soil fauna in terms of the attention paid them because of the difficulty of extracting them efficiently and identifying them. Several fungus-gnats and moth-flies live in loose soil and litter throughout their adult stages as well. Some scuttle-flies have gone still further in their commitment to an earth-bound way of life through the atrophy of wings so that they can only hop and scuttle among dead leaves and decaying vegetation.

There are three main groups of Diptera which can be distinguished in the larval as well as the adult stages. The most primitive suborder contains the midges, gnats, March flies and crane flies. The larvae of these are usually long cylindrical creatures with a complete head capsule and mandibles that move in a horizontal plane. The second group includes soldier flies, robber flies, horse flies, long-headed flies and a few other families. The larvae are also elongate with distinct, if incomplete, head capsules and sickle-shaped mandibles. They, like the adults, are largely predaceous. A great diversity of larvae are represented in the third, most advanced, group of flies, but they all have a vestigial head-capsule and hook-like mandibles which can sometimes be seen through the skin. Many of these are soft, plump maggots, essentially like the 'gentles' used by anglers as bait. Others, occurring in various kinds of decaying vegetable matter, have conspicuous spines or are strongly flattened and tough-skinned. The most bizarre form is the rat-tailed maggot of the drone fly *Eristalis tenax* whose telescopic siphon enables it to breathe air while remaining submerged in semi-liquid, anaerobic mud or dung.

As usual, most attention has been paid to pest species. The large bulb fly *Merodon equestris* and lesser bulb flies *Eumerus* species belong to the same hoverfly family as the drone fly. Their larvae attack *Narcissus* and other similar bulbs making them go rotten. As with so many pests, they become a problem only where the food plant is grown consistently in an area in monoculture, as enormous populations of the insects are then able to develop in a few generations.

Leatherjackets are the larvae of crane flies or daddy-long-legs. They occur in grassland, sports turf and corn fields, and can represent a significant force of subterranean grazers. The most frequent offender in farmland is *Tipula paludosa* which lays up to several hundred eggs during May and June. The larvae feed through the autumn, winter and spring, developing into the characteristic grey, wrinkled cylinders nearly 40mm in length before pupating. Leatherjackets are much preyed upon by rooks, starlings, gulls and lapwings which probe for them just below the surface of the soil.

The most economically important soil-living Diptera in this country belong to the genus *Delia*. The adults look like small house flies and the larvae are typical maggots. *Delia coarctata* the wheat bulb fly, *D. radicans* the cabbage root fly, *D. antiqua* the onion fly, *D. planipalpis* the turnip fly and *D. platura* the bean seed fly are pests of several important crops as their common names indicate. They share a large responsibility for the use of insecticidal sprays and seed dressings on agricultural land.

The wheat bulb fly lays its eggs on bare or sparsely covered ground in July, but these do not hatch until February when the young larvae burrow through the soil in search of young cereal plants, making use of root exudates to home in over the last few millimetres. They then bore their way into the shoot killing or damaging the young wheat plant. They are, or used to be, mainly a problem after potato crops were followed by winter wheat. Fields that were harvested later did not attract egg-laying flies, and those that were sown with spring wheat were unaffected because most larvae perished before the crop germinated.

Beetles constitute by far the largest and most diverse group among the larger soil-and ground-living arthropods. Despite their enormous variation in size and shape, adult beetles are readily recognized in almost all cases by the horny wing cases or elytra which conceal the delicate flight wings underneath. In the rove beetles the elytra are greatly shortened, and in many ground beetles the second pair of wings has atrophied, while the female glow-worm is neotenous in becoming sexually mature while retaining a larval appearance without any wings. In picking out a few with a vegetarian diet, one must recognize that these represent but a corner of the whole picture. The next section makes some redress by considering beetles as predators but this still omits dung beetles, burying beetles, scavengers and others.

Weevils probably constitute the largest family of insects in the world. They all feed on plants or plant products and are often highly specific to particular plants and parts of plants. The great majority occur on aerial parts but some species have larvae that feed on roots or other subterranean structures. The attractive, irridescent green nettle weevil is a good example. Its adult life is confined to about a month while the rest of the year is spent underground. Weevil larvae are soft, plump grubs with curved and legless bodies befiting their sedentary way of life.

Species of *Trachyphloeus* are perhaps the best examples of ground weevils for both adults and larvae live at the roots of low-growing plants. They are often cryptically coloured in greys and duns, and covered with encrustations which render them very inconspicuous. Another weevil, *Barypeithes pellucidus*, was the most common species of beetle caught in pitfall traps in a wide-ranging survey of London gardens. This was unexpected for, usually, it is the active, predatory species that stumble into such traps. It is not known to be harmful but is thought to be associated with shrubs and young trees and must, therefore, be viewed with some suspicion.

The larvae of chafer and scarab beetles are rather like overgrown weevil grubs with small legs and a swollen abdomen. The cockchafer and its close relatives are known as May or June bugs and are familiar for their habit of flying to lights at night. The larvae used to be serious pests of grassland in this country but have declined in significance as the amount of permanent pasture has decreased so markedly since the 1950s; short-term ley grassland is no use as the larvae need four years to mature. The adults constituted a very important food item for horse-shoe bats whose numbers have declined seriously in recent years.

Wireworms became a serious pest during and after the war because of the ploughing up of old pasture. Species of *Agriotes*, *Limonius* and *Athous*, in particular, could occur at high densities of the order of 2-3 million an acre, around 600 a square metre, especially on heavy loam and clay soils. They too have a four or five year larval life and when the pasture was ploughed up they turned their attentions, *faut de mieux* perhaps, to the potatoes, cereals, onions and other crops that were planted subsequently, causing damage for several successive years.

The adults of these slim, elongated larvae are known as click-beetles or skipjacks on account of the way they fall to the ground when disturbed and then leap into the air with a loud click to right themselves – and startle would-be predators. A.D Imms gives a detailed account of this behaviour and its significance in his book on Insect Natural History in this series.

Beetles as predators

The largest family of soil and ground-living beetles are the rove beetles or Staphylinidae, instantly recognizable by their waistcoat-like elytra which leave the last six segments of the abdomen exposed. They are mostly small, black and insignificant – definitely a group for specialists – feeding mainly on fungi, spores, soil algae, carrion and decaying vegetable matter. Many species, however, are carnivorous, and the best known of these is the devil's coach horse *Staphylinus olens*. It reaches a length of 20-28mm and often attracts attention through its habit of wandering about garden paths and even into houses. When disturbed, it adopts a characteristic threatening posture by opening its mandibles wide and erecting the end of its body. This suggests both an ability to give a powerful bite, which it can, and an ability to sting, which it cannot. It is, in fact, preparing to discharge a strong unpleasant odour, from vesicles at its rear end, which has a repellent effect on natural enemies. Its English name is thought to derive from a medieval legend, but its origin seems to be lost in the mists of time.

Two other related families, the Pselaphidae and Scydmaenidae, must rank among the smallest beetle predators. Between one and two millimetres in length, many of them are known to feed on mites, especially the oribatid or beetle mites described earlier. The Scydmaenidae have scimitar-like mandibles with which they grasp and hold their prey while piercing their tough shells. External digestion appears to take place through the action of digestive enzymes secreted into their prey as in the case of spiders (Some Pselaphidae are ant commensals as described later).

The dominant beetle predators belong to the family Carabidae known as ground beetles, though the German name Laufkäfer, which means running beetles, is perhaps a more apt description. They are placed with water-beetles in a separate sub-order from other beetles, the Adephaga, a Greek word meaning ravenous feeders. The genus *Carabus* itself includes some of our largest British beetles, weighing up to 600-700 mg. They offer ideal subjects for studying insect anatomy for every segment of legs, mouthparts and antennae can be clearly seen

Above: Plate 1. Section through leaf litter showing a long-legged species of oribatid or 'beetle' mite of the family Damaeidae. (Photograph J. M. Anderson)

Below right: Plate 2. Rendzina profile on chalk, under grassland uncultivated for about 30 years, Porton Down, Wiltshire. (Photograph D. F. B.)

Below left: Plate 3. A Humus-Iron Podzol from the Vale of York (Photograph R. Evans)

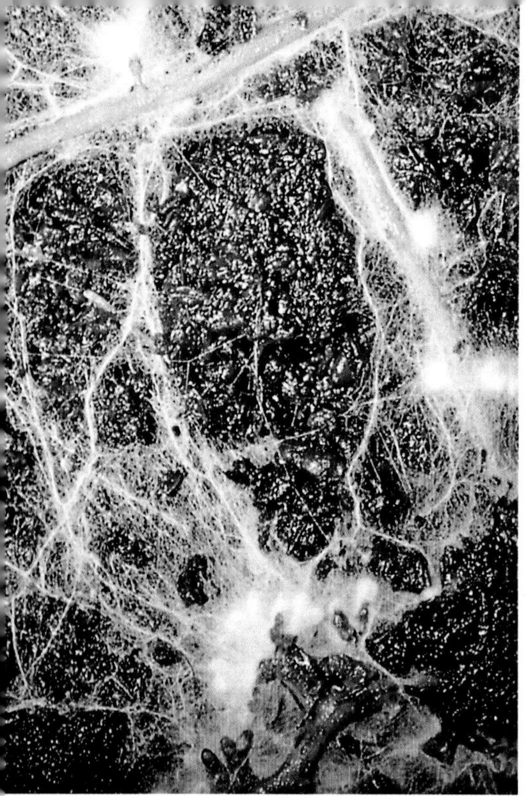

Top left: Plate 4. Ectomycorrhizal root of *Pinus contorta*. (Photograph R.D.Finlay)

Top right: Plate 5. The ghost orchid *Epipogium aphyllum* has neither leaves nor chlorophyll; it lives off its mycorrhizal fungus. (Photograph B. Dickerson)

Bottom left: Plate 6. *Neobisium muscorum* is the most common and widespread pseudoscorpion of litter and soil where it preys on springtails and other small creatures. (Photograph Dick Jones)

Bottom right: Plate 7. A soil-living centipede *Geophilus electricus*. (Photograph R. E. Jones)

Top: Plate 8. *Dysdera crocata*, whose large fangs enable it to prey on woodlice, is a common spider found by day under stones and logs. (Photograph Dick Jones)

Middle top: Plate 9. Cicada nymph *Cicadetta montana* from the New Forest. (Photograph J. A. Grant)

Middle bottom: Plate 10. The mole cricket *Gryllotalpa gryllotalpa*, a rare British insect. The first pair of fore limbs are adapted for burrowing like those of the cicada nymph. (Photograph J. A. Grant)

Bottom: Plate 11. The predatory shelled slug *Testacella haliotidea*. (Photograph A.F.Brown)

Below: Plate 12. Broadbalk fertilizer experiment (J. Stevenson, courtesy of Rothamsted Experimental Station)

Above: Plate 13. Marsh helleborine *Epipactis palustris* on Witton Lime Beds, Cheshire. (Photograph B.N.K.D.)

Right: Plate 14. The effect on young cereals of methane gas produced by decomposing buried wastes. (Photograph S. G. McRae)

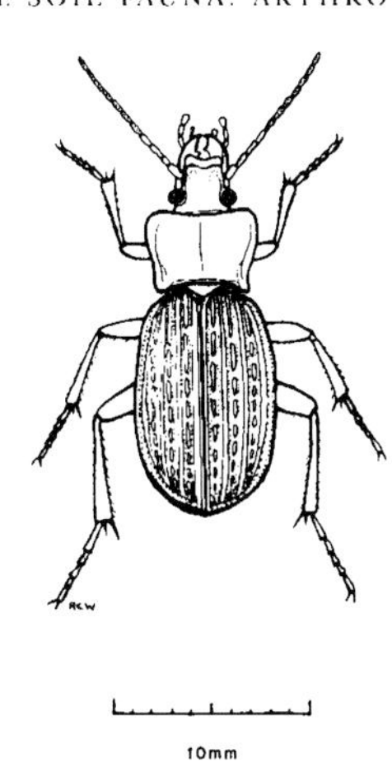

Fig. 29 *Carabus granulatus*, one of the largest ground beetles (Carabidae). (Drawn by R. C. Welch.)

with the naked eye. Different species show a wide range of elytral sculpturing and colouring, from the violet-tinted, fine matt finish of *Carabus violaceus* to the coarsely ridge-and-furrowed metallic green of *Carabus nitens*, and the tuberculate, dark copper or burnished black of *Carabus granulatus* (Fig. 29): the nineteenth century French naturalist Jean Henri Fabre described this last species as having "a cuirass magnificent with the refulgency of copper pyrites and ornamented with alternate pins and bosses".

Earthworms are thought to play an important part in the diet of *Carabus* species, especially for the subterranean larvae, but both adults and larvae will attack almost every kind of insect within reach. Much the same applies to the species of *Pterostichus*, *Agonum* and *Calathus*, except that they take smaller prey including spiders, ants, aphids, springtails and the larvae of flies and other beetles. Experiments have shown that these ground beetles will consume more than their own body weight of food in a day. More often, of course, food is a strictly limiting resource which must be continuously searched for through the litter and soil.

Whereas most of the larger ground beetles are nocturnal, the closely allied tiger beetles, belonging to the genus *Cincindela*, are active in bright sunshine. They are often common in open sandy heathland where they may be seen in early spring running rapidly over the ground and taking short flights. The prominent eyes and huge scimitar-like mandibles betray their hunting nature, while their equally voracious larvae adopt a strategy of concealment. They lie in burrows up to 30 cm deep waiting for passing prey which are seized and dragged down in a manner reminiscent of the tropical ant-lion larvae, which are related to our lace-wings, or the *Atypus* spiders described earlier.

In addition to these catholic predators, there are many which specialize on certain prey. One ground beetle which is obviously beneficial to man is *Bembidion lampros*, a brassy, diurnal species of 2.5-3.5mm common in arable soils. Here it plays an important role in consuming the eggs of flies such as the cabbage root fly *Delia radicans* which can cause much damage to brassica

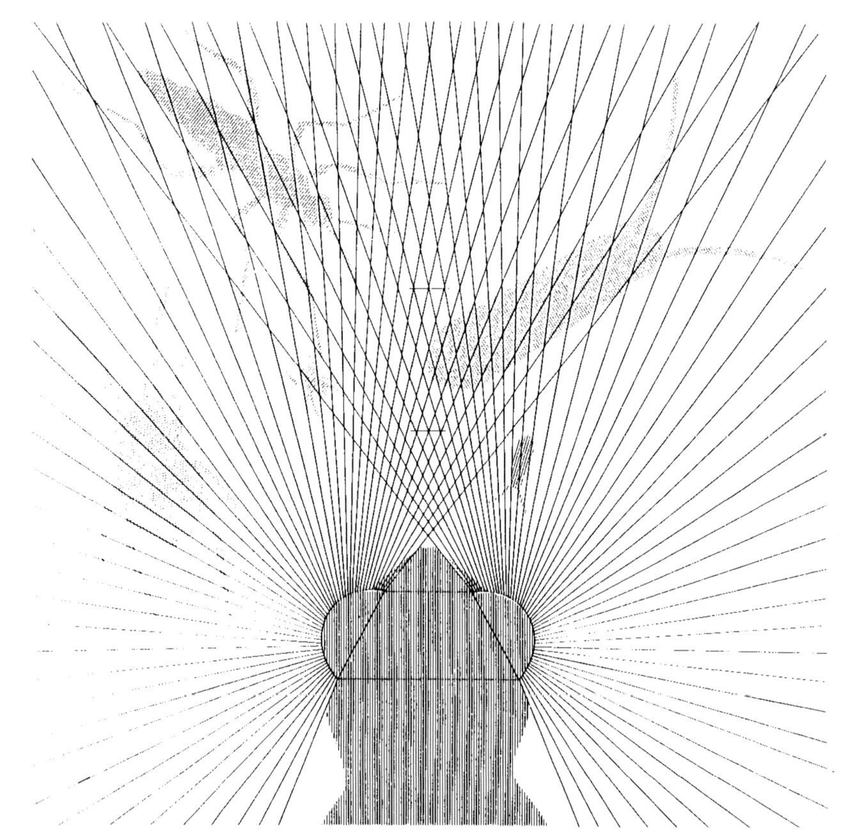

Fig. 30 Compound eyes and field of vision of the predatory ground beetle *Notiophilus biguttatus*. (From T. Bauer, 1981.)

crops. *Calasoma* species are particularly partial to the subterranean cutworm larvae of several noctuid moths mentioned earlier. *Dyschirius* pursues the burrowing rove beetle *Bledius*, using its remarkably broadened fore-legs like the mole and the mole-cricket mentioned earlier. Even snails are attacked. The flexible, Carabus-like *Cychrus* can insinuate itself into the shell, and, like the quite unrelated genus *Silpha*, can overcome the defensive slime of the snail with digestive secretions.

Modern techniques have allowed some fascinating studies to be made on the adaptations that may be needed for a predatory way of life. The two-spotted ground beetle *Notiophilus biguttatus* is a common diurnal species, about 5mm long, that hunts by sight. It has huge compound eyes each containing between 900 and 1250 separate facets. Together, these cover a visual arc of 200° with a binocular overlap of 74° within which it can gauge accurately the position and distance of its prey before it makes an attack (Fig. 30). It lives largely on springtails but the larger, surface-living kinds are hard to catch for they can spring several centimetres to safety if not grasped firmly at the first attempt. The *Notiophilus*, therefore, first lines up on its prey and then

Fig. 31 Cine film series of *Loricera pilicornis* lunging at a springtail: 1–7 antennal contact, 8 the springtail jumps, 12–15 the beetle strikes the antennae together (too late), 16–20 the antennae are reopened. Single frames from a film recording at 500 frames/second. The beetle is 7 mm long. (From T. Bauer, 1982.)

creeps forward a millimetre or less at a time with mandibles wide open until it is between 1.1 and 2.3mm away before making its lunge. An unmistakeable sign that it is ready to make an attack, usually at about 5mm distance, is when it lays its antennae back at an angle. If the springtail moves towards it before it is ready, it withdraws a bit and then advances again until it reaches the critical distance.

Its larva, like most insect larvae, has poorly developed vision, but it too must make a successful living as a hunter. This it does through the use of other senses. First, it is guided to appropriate places in the soil and litter by seeking out humidity conditions favoured by its prey. Here it is able to detect any aggregations of Collembola by chemical cues, and it starts to search using slow, sweeping movements of the head. At intervals it stops and remains motionless with head raised in wait for passing prey. If it moves out of an area where Collembola have recently been, it either turns back again or moves on quickly ur til it reaches another area pervaded by their scent. An attack is triggered by contact with special sensory hairs on the head called trichobothria. Such hairs occur widely in insects (and spiders). They usually consist of a very fine whip-like hair (flagellum) coming from a circular, thin, flexible pad which bends when the hair is moved and sends off a nerve signal. In the case of *Notiophilus*, this signal causes it to dart forward until it touches its prey

Fig. 32 A Half-grown larva of large blue butterfly *Maculinea arion* with adult ants and ant brood of *Myrmica sabuleti*. (Photograph J. A. Thomas.)

with the two sensory short horns on its upper lip, or labrum, when its jaws snap together.

These details of behaviour were watched and analysed by Thomas Bauer in Germany using a high speed camera to reveal details. He also described the attacks made by another collembolan specialist *Loricera pilicornis*. This beetle uses its eyes to hunt by day but it is also able to catch prey in total darkness with the use of its antennae. These are unusual in that the nearer segments each bear several strong bristles which project downwards and inwards and help to sweep prey towards its jaws. When *L. pilicornis* detects a springtail, visually or by antennal contact, it moves forward quickly with its antennae held about 80 degrees apart until it touches the springtail with its palps. The antennae are then struck together within 12 milliseconds and the mouthparts grasp the prey. The springtail species that Bauer studied had an average reaction time of 25 milliseconds after being touched, and often an individual managed to jump clear (Fig. 31), but when there was a cluster of them around a food source the antennal bristles swept several together at the same time and one at least was often trapped.

Ants and ant-hills

As hunter-gatherers, pastoralists, horticulturalists and artisans, the reputation and influence of ants in soil communities is unique. One cannot but be impressed by the sense of energy, organization and purpose centred on a large nest of the wood-ant *Formica rufa* in a pine wood. Several species of ants have colonies of 10,000 or more individuals but *F. rufa* colonies are reputed to reach 300,000. Many invertebrates are taken as prey by some species of ants, but a motley retinue of mites, insects and other arthropods are scavengers and commensals within ant's nests.

The most famous of these dependants, of course, is the large blue butterfly.

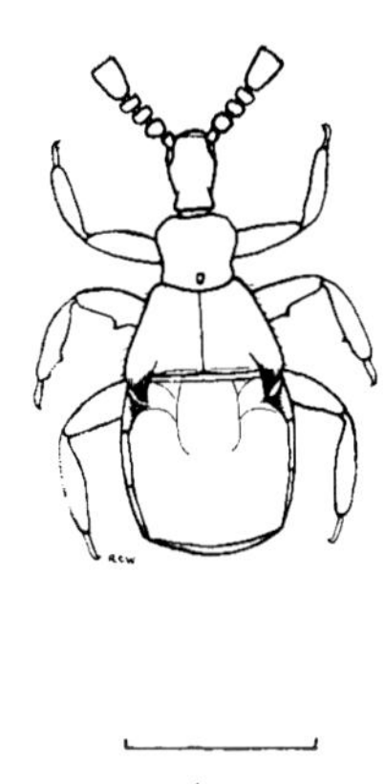

Fig. 33 The pselaphid beetle *Claviger testaceus* which lives in nests of the yellow ant *Lasius flavus*. (Drawn by R. C. Welch.)

Its larvae are taken underground by ants, and live in the nests for nine months feeding on the ant brood and producing attractive secretions in return. They can only produce these secretions in their last larval stage, and so they remain very small until they have been adopted and then grow enormously in the nests of their host (Fig. 32).The main requirements of this rare and handsome insect have been known for some time – its need for wild thyme plants for egg laying in addition to the ant colonies for larval development. Unfortunately, the extent to which the large blue larvae depended on a particular species of ant, *Myrmica sabuleti*, was not fully realized; nor that this ant depended on extremely closely grazed turf, less than three cm high. By the time research had unravelled the complex details it proved too late to halt the spiral of decline that led to the extinction of the large blue in Britain in 1979. The story shows just how subtle can be the interconnections between soils, soil animals, plants, flying insects and grazing mammals.

Very short, rabbit-or sheep-grazed turf lead to high soil temperatures which favour *Myrmica sabuleti*: slightly taller vegetation with more shading of the ground favours *M. scabrinodis*. This latter species is only one fifth as good a foster parent for caterpillars of the large blue, and so could not support a population of the butterfly within the small areas of suitable vegetation remaining in southern England. An attempt has now been made to reintroduce the Scandinavian race of this butterfly after improving the habitat for *M. sabuleti* by close grazing, and there is cause for cautious optimism.

The interesting little blind beetle *Claviger testaceus* (Fig. 33) is the most specialized of the myrmecophilous Pselaphidae, mentioned earlier. It lives in the nests of the yellow ant *Lasius flavus* where it is fed by its hosts with regurgitated food, though it will also feed on ant larvae and pupae, dead ants, flies etc. In return, it produces an attractive secretion like that of the large blue. It has been seen to cling to the bodies of queen ants, and is probably transported to new nests in this way – an example of phoresy.

The biology and ecology of British ants have been comprehensively described recently by M.V.Brian in the New Naturalist volume on ants, so this section concentrates on ant-hills. These were once a common feature of our

Table 5 The responses of some grassland plants to mounds of *Lasius flavus*

	Percentage affinity for ant-hills.
Calcareous grasslands	
More abundant on ant-hills	
Thyme-leaved sandwort *Arenaria serpyllifolia*	100
Wall speedwell *Veronica arvensis*	100
Common mouse-ear *Cerastium fontanum*	88
Common thyme *Thymus drucei*	87
moss *Bryum* spp	80
Comon rockrose *Helianthemum chamaecistus*	73
Yellow oat-grass *Trisetum flavescens*	69
Squinancy wort *Asperula cynanchica*	67
Common bent grass *Agrostis tenuis*	66
Hedge bedstraw *Galium mollugo*	65
More abundant in pasture	
Stemless thistle *Cirsium acaule*	1
Salad burnet *Sanguisorba minor*	4
Devil's-bit scabious *Succisa pratensis*	5
Glaucous sedge *Carex flacca*	6
Dropwort *Filipendula vulgaris*	11
Cat's ear *Hypochaeris radicata*	14
Common milkwort *Polygala vulgaris*	14
Rough hawkbit *Leontodon hispidus*	14
Horseshoe vetch *Hippocrepis comosa*	15
Upright brome *Bromus erectus*	16
Acidic grasslands	
More abundant on ant-hills	
moss *Polytrichum piliferum*	99
moss *Polytrichum juniperum*	98
moss *Dicranum scoparium*	86
Early hair-grass *Aira praecox*	62
Sheep's sorrel *Rumex acetosella*	61
More abundant in pasture	
Mat-grass *Nardus stricta*	3
Common tormentil *Potentilla erecta*	9
Field woodrush *Luzula campestris*	10
moss *Pseudoscleropodium purum*	12
Sweet vernal-grass *Anthoxanthum odoratum*	18
Heath grass *Sieglingia decumbens*	18

(Shortened lists from T. J. King 1977 and 1980)

landscape but their agricultural significance in grassland, before the advent of heavy machinery in the 1940s, has been largely forgotten. They posed a major obstacle to equipment but were also a valuable 'fertilizer' when spread over the ground. Sir George Cornwall's account book for Moccas Deer Park, Hereford, provides an example of the problem they were considered to be in 1884. Entries for January and March of that year itemized payments made for 'cutting ant-hills in lawn' and record a total of 4550 ant-hills destroyed. Today they are almost banished from our more intensive farmscape and must be looked for in fragments of chalk downland, isolated pastures, railway cuttings and other odd neglected corners where old grassland has persisted.

In Europe and North America several species of ants build ant-hills, but in this country they are almost always the product of the little yellow ant *Lasius flavus*. Mound building may be partly a response to climate, as in southern Europe and in high mountains the yellow ant lives under stones. It is very common in southern Britain where ant hills can reach extraordinary densities if undisturbed.

T.J.King counted ant-hills in a large number of grasslands on both chalky and acidic sandy soils using 1000 square metres as his standard sampling unit. Four counts from Roche Court Down on the Porton Ranges in Wiltshire ranged from 110 to 400 ant-hills – up to 4000 per hectare (Fig. 53, p. 134). At Aston Rowant National Nature Reserve in Oxfordshire, he surveyed 8450 square metres and counted 1947 ant-hills, while several estimates from smaller areas in Wiltshire and Sussex fell between 1200 and 3500 per hectare. The highest density was on acidic grassland in the Gower Peninsula in south Wales where he encountered 497 ant-hills in 1000 square metres, almost one in every two square metres.

Ant-hills have a definite local micro-climate caused both by their capacity to absorb warmth from the sun on their south-facing slopes and by their free-draining nature. Much of the rain runs off the surface of a mound, and that which penetrates can quickly drain away through the galleries which permeate it. Dew, frost and snow lie for longer on the north side of mounds than on the south, where, on the other hand, temperatures can often reach 40°C on bare soil on summer days. The mounds are built little by little by ants bringing up soil particles in their mandibles from a metre or more deep. These particles are glued to the existing surface of the mound and to plants, some of which continue to grow and thus provide a framework. The coarser sand grains (more than 0.5mm), are less easily transported, and so the mound comes to have a disproportionately large amount of fine soil particles compared with the surrounding soil – 84% as against 16% in the case of one study.

Not only are the physical properties distinctive but there appear to be chemical differences as well. These have been noticed in Britain, Europe and the United States. In Britain, higher quantities of extractable phosphorus have been found in some instances, while the organic matter content of the mounds is generally lower than in the soil around. Ant-hills always have a higher potassium concentration than the surrounding soil. They are often more alkaline as well, probably due to nutrient-rich clay particles brought up from lower down the profile. In Polish meadows, a similar increase in pH, phosphorus and potassium was accompanied by marked changes in the bacterial and fungal microflora. The researchers attributed the increased microbial activity to the decomposition of organic material imported into the nests. Not all chemical differences may be directly caused by ants; rabbits are known to use mounds as latrines, and one exceptionally acidic and nitrate-rich ant-hill in Colorado was deemed to have been used as a coyote 'lamp-post'!

These various changes wrought by ants are clearly reflected in the vegetation. Indeed, there is probably no other soil animal which modifies the vegetation so conspicuously in this way in temperate regions – though subterranean grazers such as root aphids and wireworms probably affect the competition between some plant species. Certain small annual plants, such as thyme-leaved sandwort and wall speedwell, and a few deep-rooted perennials, such as thyme and rock-rose, are particularly associated with ant-hills. The annuals need bare soil for germination, and they flower and set seed early in the year before drought becomes severe. The perennials must withstand periodic droughts and having soil continually heaped on them. Table 5 lists some of the plants found to be more abundant or noticeably scarce on ant-hills, compared to surrounding grassland, on chalky and acidic soils. On heathlands, the mounds may be distinctive in supporting heather. Mole-hills also provide germination niches for plants, as described later.

One of the most interesting recent discoveries about ant-hills is the relationship between their size and the age of the grassland where they occur. A multidisciplinary team of a botanist, an entomologist, a soil scientist (D. F. B.) and an historical geographer were studying the ecology of Porton Down in Wiltshire when they noticed that the average size and density of of ant-hills differed markedly in adjacent areas. Old maps and records showed that these areas had been cultivated in the past and then allowed to revert to grassland at different times (see chapter 7); those cultivated more recently had fewer and smaller ant-hills than older areas. It was known that ants continue to build their nests so long as they remain in residence, for some nests had been under continuous observation for 22 years. However, if mounds become shaded by vegetation, or are taken over by another ant species, they stop growing and even decline again. Faced with a vista of ant-hills,the problem was to decide what measurement of size or density would be the best guide to age. Was there any consistency from site to site, or from one soil type to another?

King devised an empirical formula based on the mean volume of the five largest ant-hills out of a thousand counted out in a standard area:

Age of field since last ploughing (years) = Mean volume V (litres)/1.04+5.0 where the volume of a mound was calculated from the mean height (h) and mean radius (r) by $V = \pi h\,(3r^2 + h^2)/6$

This formula seemed to be reliable for chalk grasslands between 15 and 165 years old in Wiltshire; a former golf course at Porton was predicted to be 56 years 'old', and subsequently found to have been abandoned 56 years before. It also gave surprisingly good results for other datable grasslands on alluvial soil in Surrey, clay soil in Leicestershire, limestone in Oxfordshire, and even on acid, sandy soil in Denmark. Ant-hills seem to reach their maximum size of 200 litres after about 165 years so fields with ant-hills of this size could be much older.

5

Other Soil Animals

Earthworms

To most people, earthworms are the soil animals *par excellence*. Their secretive but industrious activities have attracted the attention of naturalists from the time of Gilbert White, a century before Charles Darwin. Conspicuous surface worm casts, which are considered to be such a characteristic earthworm phenomenon, are, in fact, produced only by two species in Britain: the long worm *Aporrectodea longa* and a large form or morph ('nocturna') of the grey worm *Aporrectodea caliginosa*, which is rare. Other species eject their casts mainly within their burrows or in cracks in the soil, so an absence of visible worm casts does not necessarily mean a lack of earthworms or activity.

Darwin noticed how worm casts produced a covering of fine 'mould' over the soil surface, which resulted in the burying of stones and archaeological features such as the floors of Roman villas. He estimated that, in places, the annual production of casts amounted to some ten tons an acre (about 25,000 kg per hectare), and a very similar figure was obtained by A.C.Evans at Rothamsted in 1948. Surface casts, which are merely unsightly on lawns, can be a real nuisance on golf greens, and occasionally a problem in crops where seedlings can become buried.

Nothing can surpass Darwin's careful observations on "the formation of vegetable mould through the action of worms" in which he described the way that worms consumed, macerated and mixed vegetable matter with soil, and thereby contributed to the formation of humus. One of the features of a good, worm-worked soil is the presence of water-stable soil crumbs. These do not collapse so readily under wet conditions, and they therefore help to maintain a better structure than where there are no worms. Nowadays, we can describe this process in more biochemical, if less approachable, terms as being due to the secretion of mucopolysaccharides by earthworms with, in M.H.B.Hayes' words, "appropriate amounts of calcium to allow divalent cation bridging to take place between the humus substances and the negatively charged inorganic colloid species". The Sports Turf Research Institute has elaborate specifications for constructing golf greens which will maintain the right physical conditions while, at the same time, keeping worms at bay with pesticides.

There are about 200 native species of earthworms in Europe, all belonging to the family Lumbricidae, but only 26 occur in Britain (apart from a few imported species occasionally found in greenhouses and gardens). This impoverished fauna reflects their slow immigration rate and the relatively brief time available after the last Ice Age before the land bridge was cut by the English Channel. Other families predominate in the tropical and south temperate regions where worms sometimes reach very large size: the record is

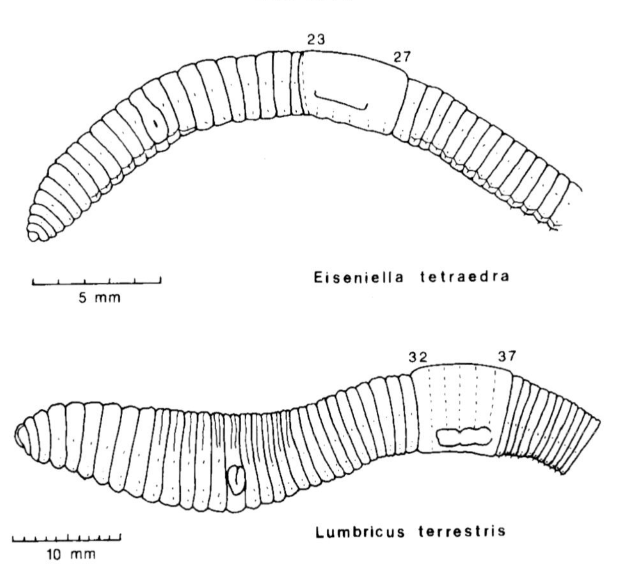

Fig. 34 The largest British earthworm, the lob worm *Lumbricus terrestris*, and one of the smallest, the square-tailed worm *Eiseniella tetraedra*. The position of the girdle, or clitellum, is an important diagnostic character in earthworms. *L. terrestris* is common in pasture and deciduous woodland soils, *E.tetraedra* occurs in ponds and soils near water. (Drawn by B.N.K.D.)

held by *Megascolides australis* of Australia which can attain a length of nearly three metres, and whose burrows go down perhaps 15 metres.

One can divide the British species of earthworms roughly into three ecological groups depending on their occurrence and burrowing behaviour. First, there are several that live close to the soil surface in accumulations of leaf litter and other vegetable matter including compost. These species are characteristic of woodland conditions, but some are also early colonists of derelict land as surface vegetation develops (Chapter 10). Then there are those that burrow in the upper soil layers and do not depend on surface organic debris. These can survive arable conditions. And thirdly, there are the deep-burrowing species already mentioned, such as the surface-casting *Aporrectodea* and the blue grey worm *Octolasion cyaneum*.

In many areas the most common large earthworm is *Lumbricus terrestris*, known as the night crawler in America and the lob worm in this country (as distinct from lug worm, its distant, marine cousin found on muddy coasts). This species can reach a length of 30 cm and a weight of 10 g (Fig. 34), making it much the largest soil animal in this country apart from the mole. It makes permanent burrows that may go down as much as three metres, but it feeds at the soil surface so, in this respect, it belongs with the group of surface-living worms. On mild, damp nights it can be seen stretched out on the ground sweeping a circular area in search of food items. Its tail remains hooked into its burrow where it is expanded into a diamond shaped anchor to ensure swift retreat if necessary. The light of a torch, or the vibration of a footstep, is enough to trigger off the contraction of its longitudinal muscles,

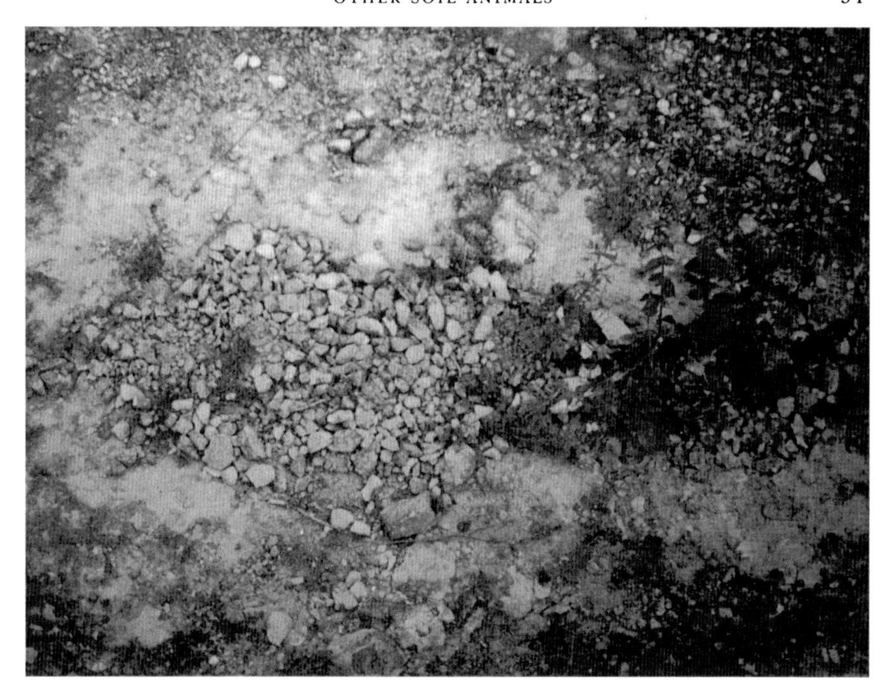

Fig. 35 Aggregation of stones by the lob worm *Lumbricus terrestris* on a quarry floor. Notice the seedlings of small toadflax *Chaenorhinum minus* among the stones and the surrounding area swept bare by the earthworm. (Photograph B.N.K.D.)

and its instant disappearance. Worms are highly nutritious, and *L. terrestris* is much sought after by badgers, foxes and owls at night as well as by thrushes, rooks, lapwings and seagulls by day.

All sorts of materials are seized and dragged into its burrow, including feathers and bits of cloth or plastic; a somewhat mindless activity, seemingly, but almost anything gathered into its midden in its original woodland habitat would eventually become edible. Experiments by J.E.Satchell have shown that *Lumbricus terrestris* actually has quite a good sense of taste for it shows clear preferences for leaves that are rich in nitrogen and sugars: hawthorn, apple, ash, alder, birch, sycamore, nettle and dog's mercury are more palatable than oak, beech, pine, spruce or larch which have distasteful tannins or polyphenols.

It is a common sight in autumn to see the ends of leaves protruding out of the soil on a path or lawn; and it was Darwin who showed unequivocally, through a large number of experiments, that a leaf was most often seized by the narrow tip, which would offer least resistance when drawn into its burrow. The same applied to elongated triangles of paper smeared with fat. This showed that the worm was guided in some way by shape not texture. The aspiring researcher would do well to read this account as a model of thorough, step-by-step discovery. Darwin did not have the benefit of statistics to help him reach 'significant' conclusions. Nevertheless, he realized the need for repeating experiments many times to be sure that the results could not be due to chance.

On bare, stony ground, such as a gravel path or weathered quarry floor, the mouth of the burrow is often marked by a small cairn of stones (Fig. 35). Some of these stones may have been pushed up from below to make more space. Others have certainly been dragged there. A colleague, Christine Brown, once witnessed, by the light of a dim street lamp, a large *Lumbricus* feeling the surface of an angular pebble with its lips. It was apparently selecting the best corner to get hold of for, after a few seconds, it fastened its mouth over this, and jerked the stone a centimetre or two towards its burrow.

The less palatable kinds of leaves need a period of weathering before they become acceptable, but nevertheless, earthworms can consume more oak and beech litter than all other soil invertebrates put together. This has been shown by burying leaf discs in nylon bags with various mesh sizes so as to exclude different groups of invertebrates according to their size (see Fig. 50). In apple orchards, *L. terrestris* can remove 90 per cent of the fallen leaves during the winter months, equivalent to 1.2 tonnes per hectare dry weight. Thus the cycle of nutrients from plant to soil is set in motion again.

What would happen if there were no worms? Frank Raw looked at an orchard from which all worms had been eliminated by residues from copper-based sprays over several years, and found that a layer of peat-like organic matter had developed in the top 1-4 cm. Dutch investigators found the same thing, and they also noticed that the underlying soil structure was itself starting to deteriorate. Where there had been a stable, subangular, blocky condition with high porosity, the soil was becoming more cloddy, less stable and less porous. The grass was more shallowly rooted as well and was more easily damaged by mowing.

These changes had no discernible effects on productivity in fertile orchard soils because the tree roots were already well established and able to exploit deeper soil layers. One may wonder, however, whether they might affect the establishment of seedlings. In less fertile soils, the effects of of worms may be much more important for tree growth. In Gisburn Forest in North Yorkshire, for example, there is a 30 year old experiment on the effects of growing oak, alder, Scots pine and Norway spruce separately and in pairs. In the spruce monoculture, the worm population was extremely low and consisted mainly of one small species *Lumbricus eiseni*. Mixtures of alder or Scots pine with spruce, on the other hand, gave a more palatable leaf litter which attracted more worms including the red worm *Lumbricus rubellus*. Not only was there a marked increase in the weight of worms per square metre here compared with the monoculture, but this was associated with an increased mineralization of nitrogen and phosphorus, and also with an improved growth of the spruce.

The highest populations of worms occur in permanent pastures with estimated weights of between 456 and 646 lb an acre – 511 to 724 kg per hectare. The often quoted comparison, attributed to the Danish observer Hansen, is that this weight exceeds that of the livestock grazing the land. However, this was thirty years ago, and the agricultural improvement of old pastures with fertilizers, and modern grazing practices, make this comparison less realistic and meaningful today.

Though there are no precise measurements of the total amount of soil passing through the bodies of earthworms, A.C.Evans at Rothamsted suggested that this could be up to 36 tons an acre a year (about 90,000 kg per ha) in grassland. At this rate, the fine soil in the top four inches of an old pasture at Rothamsted could all pass through the guts of the worms living there in 11 years. Arable fields have much lower worm populations, and the comparable calculation for a nearby field was 80 years.

A good way of seeing worm burrows and soil mixing is to set up a small chamber consisting of two glass plates held apart by strips of wood 1-2 cm thick with an elastic band around them. A few small worms, such as the grey worm *Aporrectodea caliginosa*, are placed inside, and the chamber is filled with different coloured layers of soil and put in a cool place for a week or two. This species has been used in experiments to show that tunnels can be made in densely compacted blocks of soil by ingesting soil particles, and not just by pushing the soil aside. This fact is important for it means that worms can recolonize arable soils that have been compacted by the passage of heavy machinery.

The deeply burrowing species can significantly improve drainage in certain instances. Some interesting observations were made by W. Ehlers in Germany by irrigating plots with water containing a blue dye, and then finding stained worm channels down to 180 cm. These established channels, in a silty loam soil with poor natural drainage, could absorb the equivalent of 1mm of rain a minute falling on the soil surface by allowing it to drain deeply into the profile.

Heavy rain and flooding can sometimes drive worms from their burrows. One of us (B.N.K.D.) counted 38 mature *Octolasion cyaneum* along a half mile stretch of country road one dull afternoon in early May after persistent rain. Dozens more could probably have been found on the grass verges from where they came. Similarly, large numbers of the allied *Octolasion tyrtaeum* were expelled from their burrows at Woodwalton Fen, Cambridgeshire, in July 1968 after exceptionally heavy rain when parts of the fen were flooded for several days.

In dry weather, burrowing species such as *L. terrestris* retreat to more congenial conditions deep in the ground. Species without permanent burrows curl up in a cell and go into a period of quiescence or diapause until conditions improve again. Earthworms have a remarkable ability to withstand desiccation; 70-75 percent of their water content can be lost during prolonged dry periods. Nevertheless, a long drought is as bad as a period of severe frosty weather in causing high mortality.

The extent of wet and dry conditions is thought to control the balance between green and pink forms of *Allobophora chlorotica*. Pink is genetically dominant over green since all the offspring from a green x pink cross are pink, but these offspring are wholly or partially sterile. The pink form appears to live in drier soils such as woodland, gardens and well drained pastures whereas the green form seems to favour damper habitats.

Earthworms are hermaphrodite. The smaller, surface-dwelling species produce many egg-capsules a year which hatch and grow into adults in 20–40 weeks. Their adult life may also be only a matter of a few weeks. These are the so-called 'r-selected' worms characteristic of variable or unpredictable environments, and having a potential for rapid increase when conditions are favourable. The larger and deeper-living species live in relatively predictable environments and are called 'K-selected' species. They reproduce and grow more slowly. *Lumbricus terrestris* may take 3–4 years to become adult, and individuals have been known to live up to 30 years[1].

Most species of earthworms favour neutral or slightly alkaline soils and need calcium for their digestive calciferous glands. About a dozen species commonly occur in grassland, while only a few, such as *Lumbricus eiseni* mentioned above, can tolerate the more acid conditions found in moorland, heathland and conifer woodland. The brightly banded *Eisenia fetida*, the brandling worm of anglers, likes well rotted manure heaps and can produce dense populations there. Several other species are also found in compost and leaf litter or, like the red worm *Lumbricus rubellus* and the bank worm *Dendrodrilus rubidus*, often congregate under dung pats in pasture. River banks and marshy meadows appear to be the haunts of some less common species.

Earthworms continue their fascination for scientists in many different ways; about 2500 papers on earthworms were published between 1930 and 1981. Darwin's book on earthworms was published in 1881 so, in 1981, a Darwin centenary symposium was devoted to earthworm ecology with contributions from 20 different countries – from the Philippines to Canada and the USSR. Perhaps the most recent theme of interest in a hungry world is vermiculture, to see whether worms can be cultured on various waste materials to provide food for fish, livestock, or even direct human consumption. Not only do worms have a high protein content, but analyses show that this protein is rich in essential amino acids.

Soil nematodes

Parasitic nematodes are associated with most kinds of plants and animals; man alone is host to some 50 species which, as threadworms, roundworms or hookworms, are perhaps better known than the more unobtrusive soil dwelling forms known as eelworms. Nematodes are omnipresent in soils just as they abound in the bottom ooze of oceans, lakes and rivers. Next to protozoa, they are, indeed, the most abundant soil animals, with populations up to 10 or 20 and even 30 million per square metre in the top few centimetres of pasture and woodland soils. Here they form a major component of the interstitial micro-fauna along with protozoa, rotifers, gastrotrichs and tardigrades, crawling or swimming in films of water through the labyrinth of pores.

The term eelworm is an apt description of these tiny creatures, mostly 0.5–2mm in length by 0.02-0.1mm in breadth, as they thrash back and forth in a characteristically serpentine manner in a dish of water. Like eels and earthworms, nematodes have a definite dorsal and ventral surface but they have a much simpler musculature: they cannot expand and contract like earthworms, nor can they bend from side to side as they appear to do. Their movements are almost entirely confined to the dorso-ventral plane – as in traditional pictures of the Loch Ness monster; because of this, they are forced to lie on their sides when wriggling on a flat plate (Fig. 36).

There are perhaps one to two hundred species of eelworms that one might

---

1 r and K are parameters defining the rate of increase and upper limit respectively in the logistic equation for natural population growth: $N = K/(1+e^{(a-rt)})$, where N is population size, t is time and a is a constant.

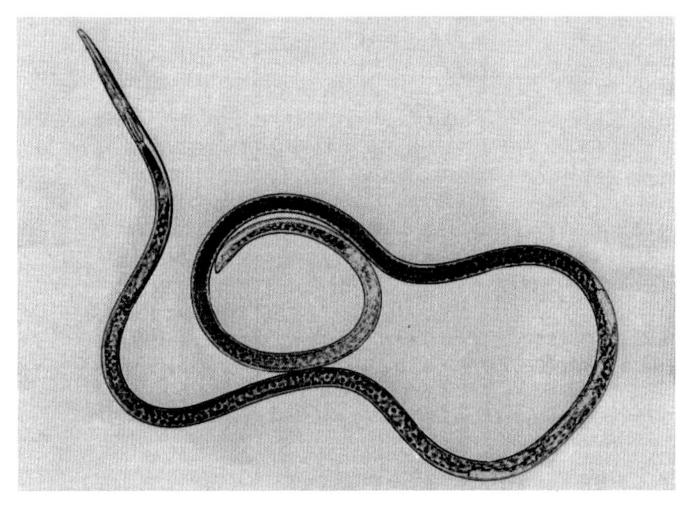

Fig. 36 The eelworm *Longidorus*; an adult female. (Photograph K. Evans.)

encounter in an uncultivated soil, but 95 per cent of these are likely to belong to ten or a dozen common species while the rest are relatively rare. Most free-living forms are very alike – vermiform, smooth-skinned and transparent – and it requires a good microscope, and much practice, to discern the characters that distinguish them. The anus is not terminal, as it is in earthworms, so there is a true tail (as in eels) whose form and relative length are useful in identification. However, the mouthparts provide some of the most important features as they immediately give clues to the method of feeding.

Microbial feeders, such as *Rhabditis*, have a narrow gullet for swallowing bacteria and algae. Predators like *Mononchus* have strong teeth in the buccal cavity, and can attack relatively large prey such as protozoa, rotifers and smaller nematodes. One individual was found to have consumed 83 larvae of a plant-parasitic eelworm in a day and 1322 in 12 weeks. A third type, typified by *Tylenchus* and *Dorylaimus*, has a sharp stylet which is used to pierce and suck the juices from fungi and roots and sometimes from other animals. The lips are pressed against the cell wall which is penetrated by repeated thrusts of the stylet. Digestive enzymes are then injected into the plant and the cell contents are sucked back.

This last mode of feeding has much in common with that of underground froghopper and cicada larvae. It has been suggested that every kind of plant has its own particular species of root nematode, and that their underground 'grazing' effect is grossly underestimated in comparison with the effects of most phytophagous insects simply because their feeding is unobserved. Add to these the predators, microbivores and omnivores and one can begin to perceive the diversity of the soil nematode fauna that might inhabit a flower-rich meadow.

The species that attack cultivated plants are best known because much work has been done to elucidate their life histories, to estimate their populations, and to find ways of controlling them. There is a complete sequence of forms from those that are fully mobile as adults – the vagrant eelworms that

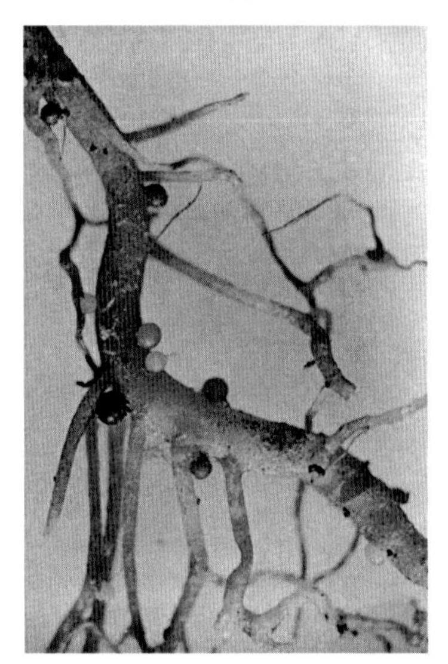

Fig. 37 Potato root eelworm *Globodera rostochiensis*; developing females and cysts on potato roots. (Photograph K. Evans.)

graze intermittently and superficially on plant tissues – to those that burrow deeply into roots and create cavities within them. The most specialized forms are found in the family Heteroderidae in which the sedentary females become swollen and flask-shaped and scarcely recognizable as eelworms at all. The typical vermiform males are sometimes rare or absent. Even when present they may be unnecessary for fertilization because, in some species, new generations are produced parthenogenetically.

The root-knot eelworms of the genus *Meloidogyne* are so called because they produce gall-like growths in the tissues they attack. The females become firmly embedded in the root, and the eggs are laid into a gelatinous sac where they can survive for over a year in the soil. They are mainly tropical species attacking important crops such as tea, coffee, peanut, tobacco and sugar-cane as well as vegetables and cereals.

The cyst-forming nematodes of the genus *Heterodera* take this life style a stage further. The eggs are retained within the body of the female which becomes a tough flask-shaped cyst after it dies. Such cysts can lie dormant in the soil for several years until the eggs are stimulated to hatch out by moisture and root secretions. The cysts are 0.3-1.0mm across and so are clearly visible to the naked eye looking like tiny earthworm cocoons.

Local infestations of these eelworms can build up in the soil until they manifest themselves as 'sick' patches in a crop, sometimes resembling nitrogen deficiency. In order to estimate their populations, detailed procedures of sampling and extracting the cysts have been devised. Because eelworms are never uniformly distributed in the soil, this involves taking a large number (500) of small soil samples across a field, mixing them thoroughly, washing a subsample over a series of sieves, and finally picking out the cysts under a microscope. Each cyst found per 500 gram sample is equivalent to a million per acre and each may contain several hundred eggs. Ten eggs per gram can cause noticeable crop symptoms, and 100 eggs per gram could produce total crop failure.

The main agricultural pests in Britain include the beet eelworm and the potato root, cereal root, pea root and carrot root eelworms. Each of these will attack several related crops, and often wild plants as well. For example, the beet eelworm is known to attack more than 30 species including mangold and spinach in the beet family (Chenopodiaceae), turnip and cress in the cabbage family (Cruciferae) and also docks (Polygonaceae) and chickweed (Caryophyllaceae). The potato root eelworm *Globodera rostochiensis* is now placed in a different genus from the other British species because its cysts are spherical rather than pear-shaped (Fig. 37).

Feeding by parasitic nematodes often produces a reaction in plant tissues which is not matched by any other invertebrate group. Some of the dividing cell walls dissolve away creating 'giant' cells with multiple nuclei enclosed within thickened outer cell walls. These giant cells may extend for several millimetres along a root. A heavy infestation interferes with normal root system development and disrupts the transport of water and nutrients, so producing the symptoms of wilting and mineral deficiency mentioned earlier. Less severe lesions can nevertheless be harmful by opening the way for secondary infections by fungi and bacteria. Some eelworms are known to be carriers of plant viruses. Species of *Longidorus* (Fig. 36), for example, transmit raspberry ringspot virus to various soft fruit, and tomato black ring virus to onion, potato and sugar-beet. In this respect they are similar to the aphids that transmit yellow dwarf virus of barley, and to the 'big bud' mites which transmit reversion disease of black currants.

The elimination of parasitic eelworms by chemical methods is difficult and involves the use of powerful soil sterilants. However, a large measure of control has been achieved since the 1960s through a combination of cultural methods, especially crop rotation, and by the use of resistant varieties of potatoes and other susceptible crops. Crop resistance can take several forms: the varieties may not be invaded by eelworm larvae at all, or else the larvae fail to develop normally. Biologically, the most interesting way of frustrating nature is that in which the plant allows only male larvae to develop fully!

Snails and slugs

There are nineteen families of land snails in the British Isles and four families of slugs comprising some 87 and 30 species respectively. Of these, fewer than a dozen are familiar to the layman, and then mainly because of their unwelcome attentions to crops and vegetables. If, however, one were to examine carefully a few shovels of litter and soil from a beech woodland or limestone grassland, shaken through a sieve and spread out over a tray, one would discover a great variety of small snails, "inexhaustibly variegated in their marblings and convolutions" to quote Teilhard de Chardin. Nor should it be assumed that these are just young individuals; most snails are less than a centimetre along the major shell axis, and the ubiquitous *Carychium minimum* is only 1.9mm in height when fully grown.

There are about 20 species that are restricted to calcareous soils, notably the chrysalis snail *Abida secale* which frequents dry, stony ground in the Cotswolds and elsewhere in south and west England, and the burrowing *Pomatias elegans* which lives in loose soil and moist leaf mould in chalk and limestone

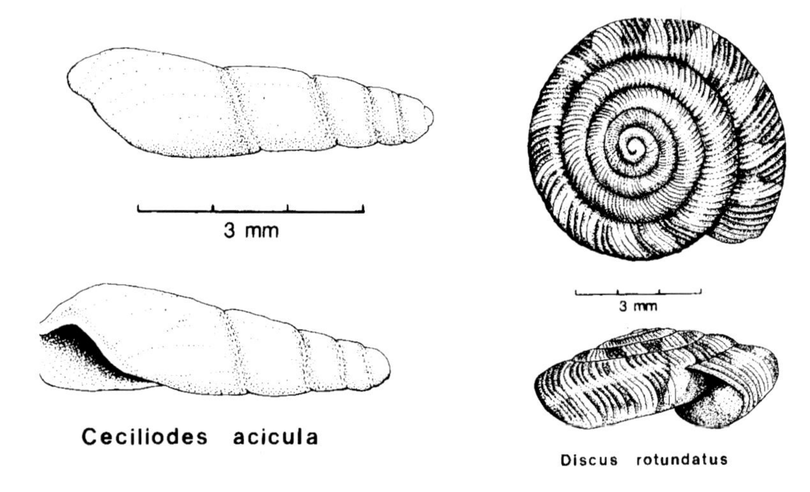

Fig. 38 Two snails of soil and litter showing contrasting forms; the blind snail *Cecilioides acicula* and the round snail *Discus rotundatus*. (Drawn by S. V. Green.)

districts. In this country, the Roman or edible snail *Helix pomatia* is confined to hillsides, banks and quarries on calcareous soils. It was almost certainly introduced by man as a culinary delicacy as it is absent from Pleistocene fossil deposits, and the earliest record is from a 4th century Roman excavation. This species must hold the record for longevity as well as size. G. Lundqvist, in Sweden, kept notes on one particular individual in his garden for 30 years, and recorded a weight of 67 grammes – as much as a large hen's egg.

Apart from variations in absolute size, the identification of snails depends largely on the ratio of shell height to width, the shape of the mouth, the number, shape, colour and texture of whorls and whether they are dextrally or sinistrally wound; a shell is said to be dextral when the coils are clockwise as seen looking down on the apex. The commonest of the door snails *Clausilia bidentata* represents one extreme form with its narrow spire-shaped shell bearing fine chasings on its 12-13 sinestral whorls. At the other extreme, the rounded snail *Discus rotundatus* (Fig. 38) is more than twice as wide as high and the whorls are dextrally coiled.

Identifying snails by their shell characteristics is not at all the same thing as classifying them into genera and families. Fossils show that some superficially similar species have descended from quite dissimilar ancestors through convergent evolution. A natural classification therefore depends primarily on internal anatomical features, especially of the reproductive system.

Slugs are a good example of a group which are superficially alike but which are highly polyphyletic in origin. They are all evolved from shelled ancestors but the British species are considered to have descended from three distinct lineages one of which has subdivided more recently. The loss of the protective shell is balanced by other gains such as the ability to adopt a very streamlined form which enables them to penetrate narrow channels and crevices in the soil. Some species are very largely subterranean notably the three species of *Testacella* (Plate 11). These can sometimes be found crawling freely on the surface at night, and can easily be recognized by the small external shell, resembling a tiny mussel shell, perched on the rear end. In all the other slugs the shell is reduced to a small plate or mass of granules embedded within the body mantle at the front end of the body.

Colour variations in slugs are common among adults as well as between juveniles and adults. Too much reliance on external characters, coupled with incomplete or inaccurate descriptions and on dissections of specimens pickled in alcohol, has led to much confusion in the past. Kerney and Cameron's field guide in 1979 included four species not recognized in H.E.Quick's monograph published in 1961, and four more species have been added since then. It was amateur collectors rather than professional biologists who first noticed behavioural differences between apparently similar slugs, especially during breeding. Morphological differences were then found in the reproductive organs of freshly killed specimens. This increase from 23 to 30 species in a group of quite large soil animals is an interesting example of the significant additions that may still be made through careful observation without sophisticated equipment.

One of these 'new' species, the Durham *Arion* in Eversham and Jackson's 1982 key, has, at the time of writing, only just received its Latin epithet *flagellus*. This species is rather similar externally to the true but apparently less common *Arion lusitanicus* so many of the earlier records of *lusitanicus* are incorrect. Similarly, the once 'easily recognized' and commonly recorded garden slug *A. hortensis*, with its bright orange or yellow sole, is actually composed of three species *A. hortensis*, *A. owenii* and *A. distinctus*, of which the last is by far the most common. Confusion over two or three species of slugs may be thought a somewhat academic problem, but it could lead to erroneous conclusions in experiments on slug control.

Most snails and slugs are hermaphrodite, like earthworms, so mating usually results in both partners laying eggs, often in batches under logs or in pockets excavated in the soil. Some species are protandrous, that is they behave first as males and later in life as females. A few slugs are self fertilizing and can therefore reproduce without having to find a mate. Some populations of the marsh slug *Deroceras laeve* are parthenogenetic, the eggs being produced without even the normal exchange of chromosomes that occurs in self fertilization. Parthenogenesis is quite common in earthworms and eelworms and widespread among arthropods including mites, millipedes and aphids, but this was the first record in molluscs.

Most snails and slugs feed on dead and decaying plant and animal matter including dung. Fungi are much favoured by some species while others rasp algae and lichens from rocks in a manner very like limpets, their marine relatives. The glass snails *Vitrina pellucida*, *Oxychilus draparnaudi* and *O. cellarius* are carnivorous, partly on enchytraeid worms and small earthworms among litter and moss. The *Testacella* slugs mentioned earlier are also predatory. They can reach 100-120mm when fully extended, and can tackle quite large earthworms such as *Lumbricus terrestris*, though they will also tackle centipedes and other slugs.

A few species of slugs give the whole group a bad name among farmers and gardeners. These are not the really big slugs such as the great black slug *Arion ater*. This may look imposing when extended to its full 150mm but it rarely frequents arable land. Rather it is its smaller relatives in the *Arion*

Table 6 Ranking of potato varieties according to their susceptibility to attack by slugs. (From N.W.Runham & P.J.Hunter 1970).

Very High Susceptibility	*High Susceptibility*	*Medium Susceptibility*	*Low Susceptibility*
Maris Piper	King Edward	Majestic	Pentland Falcon
Ulster Glade	Record	Pentland Dell	
	Pentland Crown		

distinctus group that attack various root crops, causing damage that makes them unsaleable if not inedible. They can be quite discerning in their preferences for different potato varieties if given the choice (Table 6).

The most common British slug, the grey field slug *Deroceras reticulatum*, is an important pest of winter wheat because it hollows out the newly sown grain and also grazes on the young shoots both below and above the soil surface. In the late 1960s, slug damage was held responsible for the loss of 25,000 acres of wheat (10,000 ha) in an average year. Crops of barley and oats are sometimes affected, and even newly sown leys. Damage by slugs, and other soil pests, becomes increasingly important as modern farming techniques reduce sowing rates for crops like sugar beet. Formerly, many more seeds were sown than were needed and the excess seedlings were hoed away. With monosperm seed it is now possible to sow only what is needed to grow to full size so there are fewer to spare for rook or slug.

The keeled *Milax* slugs are very largely subterranean and feed almost entirely on roots. They can burrow through compacted soil by ingesting it, like *Aporrectodea* earthworms. *M. budapestensis* is a serious pest of root crops but *M. rusticus*, a handsome pink species with black spots, lives in woodland. It was first discovered in Britain in 1986 in a wood in Kent.

Moles

The natural history of the mole has been described in detail in the New Naturalist monograph series by Kenneth Mellanby. However, one cannot conclude this chapter without briefly mentioning something of its effects upon soils. At times, its burrowing activities have been sufficiently troublesome to warrant large scale trapping simply because of the physical disturbance caused by the runs and mole-hills. A feature article in *The Guardian* in July 1983 described a mole-catcher in the northern Pennines whose record book showed a total catch of 10,000 moles during one 30-month stint in the late 1950s. "And when they sacked him, the mole heaps multiplied until there was no place for even a sheep to lie down. So at the age of 71 he finds himself back on the job."

Mole runs act like a continuous pitfall trap for soil invertebrates which drop in as they themselves burrow through the soil. Earthworms are the preferred and most important food but various insects are also taken, particularly the active larvae of beetles, flies and moths, such as wireworms, cockchafers, ground beetles, March flies, leatherjackets and cutworms. Moles also forage for inactive prey such as moth and ant pupae. Young moles may live on the surface for a time making superficial runs through vegetation.

Mole-hills and other 'molesigns' provide a rough guide to the presence of moles but tell one little about their movements. For this purpose, moles have been marked with identification rings placed around the base of the tail. This

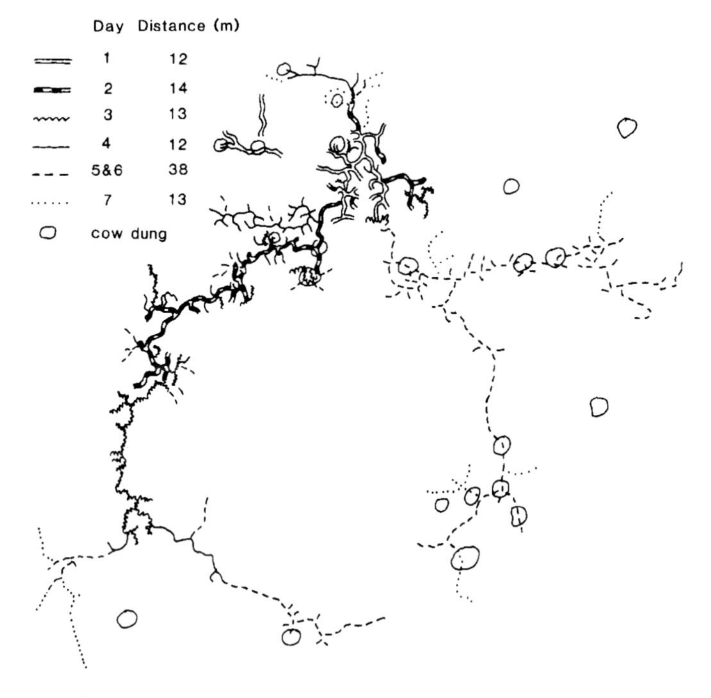

Fig. 39 Newly formed burrow system of a mole introduced into a meadow in the Dutch polders. (Redrawn from J. Haeck 1969.)

technique is particularly useful for mapping the territories of several moles in a field by recapturing them alive in traps. However, it requires an array of traps suitably placed, and assumes that you are successful in catching moles repeatedly without injuring them. A more 'high-tech' method for studying an individual is to use a ring of radioactive cobalt 60 or silver 100, and to follow its subterranean movements for periods of a few hours at a time using a Geiger counter.

J. Haeck mapped the construction of a burrow system of a mole, introduced into an unoccupied meadow in one of the Dutch polders, over the course of seven days (it disappeared after this). The mole dug a total distance of 102 metres in the course of a week at between 12 and 19 metres a day. Notice how, in Figure 39, the runs often went under cow dung which, presumably, the mole could smell some distance away, and which would attract a concentration of worms and coprophilous insects.

Once a burrow system is established, however, extension will only be necessary from time to time. In dry or freezing weather, when surface-living worms become inactive, and burrowing worms retreat to deeper layers, moles will often deepen their burrows and throw up more mole-hills to dispose of the displaced soil. Mellanby shows a photograph of a mole-hill forced up through a covering of snow. In light soils, too, the burrows collapse more readily than in loamy soils, and need more frequent reconstruction. This is particularly the case in heathland or moorland soils where earthworms and

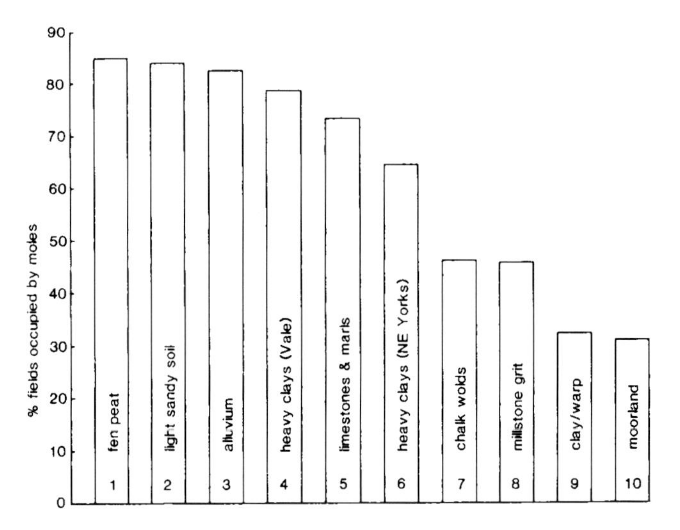

Fig. 40 Occurrence of moles in 652 pastures from districts in Yorkshire with distinctive soils. 1, Fen peat, Vale of Pickering (40 pastures); 2, Light sandy soil, Vale of Pickering (38); 3, Alluvium, Vale of York (29); 4, Heavy clays, Rydale and Vale of Pickering (52); 5, Soils overlying magnesian limestones and marls, Doncaster area (15); 6, Heavy clays over Jurassic sandstones, North East Yorkshire (222); 7, Chalk Wolds (69); 8, Millstone grit soils, South and West Yorkshire (87); 9, Clay/warp soils subject to seasonal flooding, between Rivers Went and Don, Doncaster (74); 10, 'Improved' moorland soils on southern edge of North Yorkshire Moors (74); (Data from C. A. Howes.)

other large invertebrates are sparse so that a large network is needed to harvest enough food.

The distribution and population density of moles depend on several factors, for example, food availability, disturbance and soil compaction. Mellanby described the occurrence of moles in different habitats such as farmland, deciduous and coniferous woodland, Fenland and Breckland. In this last habitat, he mentions the relatively low population occurring where 'an immense amount of surface activity gives the erroneous impression that moles abound'. Mole signs are sometimes an indication more of activity than density.

By concentrating on farmland alone, C.A.Howes has looked specifically at the importance of soil type on mole frequency. He did this by recording molesigns during journeys in Yorkshire and Humberside which passed exclusively through areas covered by peat, sand, boulder clay, alluvium and warp overlying distinct geological formations. Altogether, he made 34 transects and recorded 1165 roadside fields of which 652 were pasture. The histogram in Figure 40 shows that fen peat soils topped the frequency score with 85 per cent mole occurrence, but sandy soils in the Vale of Pickering, and alluvial soils in the Vale of York, were scarcely less good. All of these were clearly much preferred to shallow soils overlying chalk, pastures subject to regular flooding, and recently reclaimed podzols on the edge of the South Yorks Moors.

Moles live to be about five years old. Males and females occur in roughly equal numbers and have one litter a year containing about four young. For a population in an area to remain stable, two- thirds must either die or emigrate to pastures new before the next breeding season in May. Usually the young are forced out of existing runs, and it is then that one may notice new runs – for example in one's lawn. There are always some empty niches and unexploited areas for a few moles to find but the carrying capacity within a parish, say, normally remains fairly constant. Population density can only increase when food resources increase significantly. This could happen if arable land, which supports rather low populations, is converted to grassland or woodland under the agricultural 'Set-Aside' scheme in the future.

When the Dutch polders were created in the IJsselmeer, there was a dramatic population increase, and a unique opportunity to study the spread of moles into virgin land. This spread has been described in detail by J.Haeck, and some of the salient features are given by Mellanby. The surrounding banks, or dikes, were constructed and sown with grass and clover a few years before the rest of the polder was laid out. These, therefore, provided the initial bridgehead for colonization from the mainland. Because of their linear nature, they allowed measurements to be made of the rate of spread. In the Nord-ost polder, this appeared to be two to three kilometres a year on average – 20 to 30 week's digging at the rate Haeck actually measured. From the dikes, moles spread gradually into the polders along the sides of canals, ditches and roads. Pastures were often colonized first along fence lines where the soil was less compacted.

It is interesting to compare this colonization by moles with the spread of earthworms and ground beetles which other researchers studied in the polders. Worms, or worm egg-capsules, were almost certainly introduced on the roots of trees and shrubs widely planted along roadsides and around farmsteads. Their subsequent natural dispersal in raw soil seemed to be very slow. Sampling in areas around colonies inoculated into sown grassland suggested a spread of 4-9 metres a year depending on the species. This slow spread was probably due to the low organic matter and unworked structure in these raw soils. Subsequently, worm-worked soils could be seen in infra-red and false colour aerial photographs which detected the incorporation of surface organic matter into the soil.

Insects, by contrast, were able to colonize the polders very quickly by flying in from agricultural areas on the mainland. Sixty-seven species of ground beetles were caught in traps in the southern-most polder, Zuidelijk Flevoland, within a year of its being finally drained. A dozen species were able to overwinter and breed in the newly laid out road verges so there would have been a well developed soil insect fauna long before moles invaded an area.

Mole-hills have significant effects on vegetation but not in quite the same way as ant-hills. Mole-hills are a sudden extrusion of bare soil breaking the otherwise continuous plant cover. The soil may differ somewhat from the topsoil if it has come from some depth, but the texture has not been altered as it has in ant-hills. Fresh mole-hills therefore offer sites for seeds to germinate without competition from established plants, and so allow short-lived plants to maintain themselves in a sward.

A.S.Watt's long-term studies on ungrazed grassland on Foxhole Heath in Norfolk showed that there was a cycle of change, and that this cycle was caused by mole-hills. Watt recognized four stages in the cycle which he called 'pioneer', 'building', 'mature' and 'degenerate' phases. During the succession through these phases, it was apparent that eroded mole-hills were dominated first by annual species, and then by perennial herbs, before entering the mature phase with relatively few species, mainly perennial grasses. The degenerate litter-covered phase of dead grasses was colonized, not by annuals of the pioneer phase, but mostly from neighbouring plants – unless the cycle was restarted through the action of moles.

This plant community at Foxhole Heath included five of the rarer Breckland species: Spanish catchfly *Silene otites* and spiked speedwell *Veronica spicata* characteristic of the building phase, and sickle medick *Medicago falcata*, Bohmer's cat's-tail *Phleum phleoides* and the sedge *Carex ericetorum*, characteristic of the mature phase. Their relative frequency, if not their very presence, was largely dependent upon moles.

Moles, like ants, also provide niches for subterranean invertebrates. Several species are found in rabbit and small rodent burrows also, but at least ten species of beetles are considered to be specifically associated with mole nests and runs. W.J.Fordham and R.C.Welch have independently recorded the following: *Aleochara spadicea*, *Hister marginatus*, *Quedius nigrocoeruleus*, *Q. longicornis*, *Q. othinensis*, *Oxypoda longipes*, *Pycnota paradoxa*, *Eccoptomera nigra* and *Medon castaneus*. None has attracted wide enough attention to gain a common English name.

6

Bacteria and Other Microorganisms of the Soil

All fertile soils harbour a vast range of microscopic organisms which include bacteria, micro-fungi, actinomycetes, yeasts, algae, myxomycetes, myxobacteria, protozoa and others, not to mention the large numbers of small soil animals. There is a bewildering array of species and genera, some of them present in great numbers, often many millions in a gram of soil. Soil, of course, is a particularly favourable habitat, with a variety of nutrients, for this world of microorganisms.

Until about the middle of the seventeenth century, the realm of microorganisms was virtually unexplored. But at that time, in the town of Delft in Holland, Antony van Leeuwenhoek (1632–1723) was busily engaged in grinding and polishing tiny glass lenses which he mounted and assembled into compound microscopes; with these somewhat primitive instruments, he became an enthusiastic observer of many kinds of natural microscopic phenomena. It seems likely that he was able to achieve magnifications of several hundred times and so was able to detect tiny, motile living creatures in, for example, drops of dirty canal water, infusions of pepper grains, scrapings from his teeth and other such preparations. Today, we realize that the tiny animalcules that he saw were indeed living motile bacteria. He reported many of his observations to the Royal Society of London, in a series of letters which are preserved in the Society's Archives, but he was never willing to allow others access to his remarkable microscopes.

Little more than a century ago, our knowledge of the multitude of microorganisms in soil or indeed elsewhere was almost negligible. Since then, however, especially during the last eighty years or so, impressive advances have been made in scientific microscopy, culminating in the astonishing transmission and scanning electron microscopes of today. These modern instruments can achieve very high magnifications, up to a hundred thousand times or more, and so make possible detailed observations of the fine structure of bacteria or even the much smaller virus particles and similar objects.

During the nineteenth century, the majority of microbiologists, for example, Pasteur, Koch, Cohn, Lister and others, directed their attention to the study of disease-causing microbes. The isolation and identification of these pathogenic bacteria was a vital initial step towards the eventual treatment and control of such diseases.

Apart from investigations dealing with yeasts, the operative organisms of bread-making and alcoholic fermentation, and moulds and fungi, which cause rotting and spoilage of crops or materials, the existence and activities of microorganisms in soil received little or no attention. Naturally, it is understandable that serious diseases of man, plants and animals should have been

Table 7 Historical perspective of soil microbiology.

1676	Leeuwenhoek first discovered and described bacteria.
1769	Spallanzani made experiments disproving theories of spontaneous generation.
1835-37	Schwann and others discovered yeasts and their budding processes: microorganisms were recognized as the causes of putrefaction and fermentations.
1838	Boussingault showed that the nitrogen content of the soil increased on growing clovers. This did not happen in sterile sand cultures.
1853	De Bary demonstrated the fungal nature of the cereal diseases, rust and blight.
1857	Pasteur recognized the causative organism of lactic fermentation.
1861	Pasteur distinguished between aerobic and anaerobic organisms.
1872	Cohn devised a classification of bacteria.
1876	Koch identified the germ of anthrax.
1985-86	Hellriegel and Willfarth recognized the connection between nitrogen fixation and the nodules on legumes.
1888	Beijerinck grew nodule bacteria in pure culture.
1890-91	Winogradsky isolated autotrophic nitrification bacteria.
1901	Beijerinck isolated the free-living, nitrogen fixing organism, *Azotobacter*.
1904	Hiltner introduced the term 'rhizosphere' for the space with close interactions between plant roots and microorganisms.
1918	Conn devised a technique for direct soil microscopy.
1924	Winogradsky, on the basis of direct microscopic observation of soils, advanced the concept of autochthonous and zymogenous soil bacteria.
1940	Lochhead suggested grouping soil microorganisms according to their nutritional requirements.
1943	Waksman reported the discovery of streptomycin

the main focus of investigations, and that, inevitably, microbes in general became regarded as harmful. Nevertheless, there are also beneficial microorganisms and many of these occur in soil. During the last quarter of the nineteenth and early years of the present century, perhaps the most outstanding scientists who studied soil microorganisms were Sergei Winogradsky who was active first in Zürich, then in St. Petersburg and later in Paris, and Martinus W. Beijerinck who worked in Delft in Holland. These two scientists and their co-workers made many important contributions to soil microbiology. Among other problems, Beijerinck investigated the kinds of bacteria that can fix the inert nitrogen gas of the atmosphere, converting it into compounds of nitrogen which in turn become plant nutrients. Winogradsky made many important studies including investigation of the special bacteria that mediate the oxidation of ammonia to nitrates, that is to say, the nitrification process. Table 7 gives a historical summary of discoveries in soil microbiology.

In its top 15 cm layer, a typical arable soil may contain approximately two to three tonnes (wet weight) of bacterial biomass; equivalent to the weight of some 40 to 60 sheep. If we suppose that the soil contains, say, three per cent of organic matter, then about one per cent of the latter is made up of microorganisms. Although one gram of soil may contain hundreds of millions of bacteria, because of their minute size, these constitute only a fraction of the total weight of the soil. By reason of their manifold metabolic activities, however, microorganisms represent a significant proportion of the biomass of the soil. The total biomass of soil, of course, is the weight of all the living organisms to be found therein.

The detailed scientific investigation of microorganisms is for the specialist, but some indication of what microorganisms are like, and how they are isolated and identified, is essential to a full appreciation of the living soil. Bacteria are too small to be visible to the naked eye and, even with the help of the microscope, special techniques are required for their observation. Under the ordinary transmitted light microscope (i.e. bright field), bacteria cannot be seen unless first stained by suitable dyes, or rendered opaque by treatment with certain chemicals. By staining, the organisms are killed. Unstained, living microorganisms, however, may be observed directly by the use of either a dark background microscope or a phase-contrast microscope. In a dark, dusty room, tiny dust particles become visible if a beam of light is shone across the room and, as the dust particles reflect the light, they are revealed as tiny specks of light. Similarly, unstained living or dead bacteria can be seen as bright objects under the high power of the dark background microscope. Light is stopped from entering the tube of the microscope from below and the objects are illuminated instead from the side. Bacteria or other microorganisms reflect the light falling on them and become revealed as bright objects against a dark background. This is often a beautiful sight, especially when motile organisms are seen clearly swimming across the field of view. Spiral-shaped organisms, such as, spirochaetes, also present a dramatic picture under these conditions; the morphology of these and other microbes can be readily observed in this way.

Living microorganisms can also be observed, without staining, by the phase contrast microscope provided that the refractive index of the specimen differs sufficiently from that of the surrounding medium. In this technique, the specimen is illuminated by a hollow cone of light in which the phase of the light waves coming through the objective lens is shifted by a quarter of a wavelength by a device known as a phase ring. This results in interference since light passing through the slide unretarded is seen as normal white light whereas light passing through the specimen has a longer light path and arrives in the eyepiece out of phase. Usually, the image appears dark on a light background. It is possible to take photographs with all these systems of microscopy.

The electron microscope uses an electron beam instead of a light beam and it is focussed by lenses, which are either electromagnets or electrically charged plates, to form a highly magnified electron image. The resolving power is very much greater than that of the light microscope. (The light microscope can resolve objects 0.2 microns in size, whereas the electron microscope can resolve even those only 0.001 microns in size; one micron = one thousandth of a millimetre). As in a television tube, the electron beam must be produced and focussed in a high vacuum because electrons are

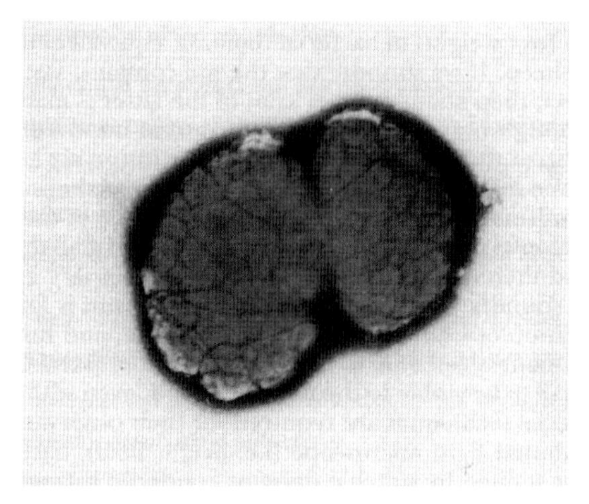

Fig. 41 Two cells of a *Nitrosolobus* sp. in division; a nitrifying bacterium isolated from a Sri Lanka tea soil. (Electron microscope photograph by N.W., magnification x 50,000.)

readily stopped even by gas molecules, and the image appears on a specially coated fluorescent screen. An electron beam is produced by heating a tungsten wire electrically near to a positively charged metal plate, i.e. the anode. If a hole is bored in the anode, electrons can pass through it as a beam. There is a third, negatively charged, electrode which, by repelling electrons serves to concentrate them into a beam. The electrical or magnetic fields in electron lenses may be varied by changing the voltage applied. Magnetic lenses are generally used and they have a minimum focal length of less than one millimetre. A typical electron microscope has two condenser lenses, an intermediate lens and a projector lens which projects the final positive image on to the fluorescent screen. Alternatively the image can be received on a photographic plate to form a photograph of the magnified object (Fig. 41).

Objects, such as, bacteria, must be supported on a collodion-coated fine metal grid (usually copper) which is inserted into the microscope. Air is pumped out to produce a high vacuum and the image can be seen on the fluorescent screen. To achieve sufficient contrast in the outlines of a bacterium, the specimen is shadowed by sputtering with heavy metal (lead, chromium or gold) heated electrically in a high vacuum. Alternatively, the specimen may be negatively stained with phosphotungstate. To see the internal structure of a bacterium, very thin sections of the organism must be prepared. The material is embedded in a hard polymerized plastic and sections are cut using an ultramicrotome equipped with a diamond knife or even a sliver of plate glass. Heavy metals are used in stains because of their capacity for scattering electrons. The electron microscope is a valuable aid in studying the fine structure of microorganisms and viruses.

To examine microorganisms with the ordinary light microscope, a variety of staining procedures is used, but these need not concern us here. The appearance and morphology of different microorganisms and their various properties and characteristics are important in their identification. Motile bacteria owe their ability to swim about to flagella, which are hair-like appendages so fine that they can only be seen under the microscope after

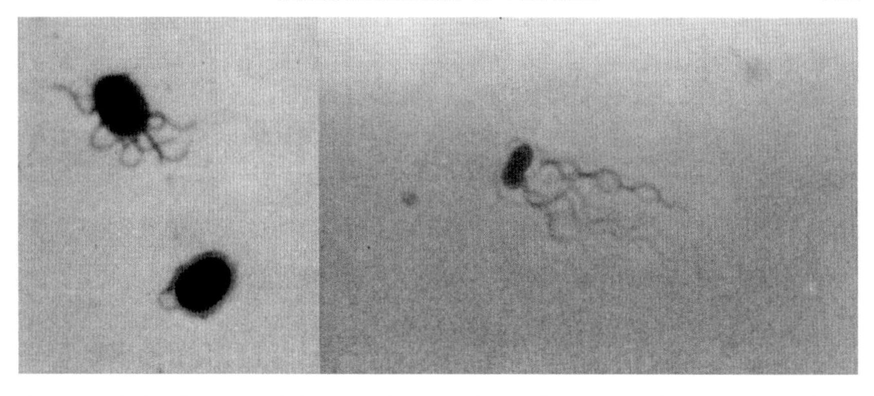

Fig. 42 Left A salicyclic acid-decomposing *Azotobacter chroococcum*, showing typical flagella. (Photomicrograph by N. W., magnification ca. 1500). *Right* A nitrifying bacterium *Nitrosolobus* sp. showing typical peritrichous flagella, isolated from a Ghana soil. (Flagella stain preparation by N. W. magnification ca. 1200.)

special treatments and staining processes; flagella can also be readily observed with the electron microscope. The number and arrangement of flagella are important characters in the identification of bacteria. Figure 42 shows twophotographs of motile flagellated bacteria.

Another type of instrument, the scanning electron microscope is especially employed to observe the surface structure of specimens, after coating them with a thin film of a heavy metal, usually gold. It is useful for viewing surface details of relatively large organisms such as micro-arthropods as in Figures 19–22.

Bacteria will multiply in a wide variety of media, that is to say, solutions of suitable nutrients. By adding gelatin or agar to such media they are converted into a firm jelly or so-called 'solid medium'. Bacteria can be inoculated on to the surface of such solidified media and on keeping at appropriate temperatures they multiply and develop into colonies of organisms which are generally circular in shape and may be several millimetres in diameter. Colonies contain thousands of millions of individual bacterial cells; the size, colour and texture of colonies are useful identifying features. If the number of live bacteria seeded on to the surface of a solid medium is suitably small, the number of colonies which develop after the appropriate incubation can be readily counted. If we assume that each colony developed from one bacterium, then the number of colonies corresponds to the number of bacteria inoculated initially on to the medium. This is the plate culture or plating method for counting microorganisms; it is also used to grow cultures of bacteria or other organisms. Moreover, it is the method *par excellence* for the isolation of pure cultures as well as for detecting contaminating microbes in presumed pure cultures, because different microorganisms yield colonies usually differing in some feature or other (Fig. 43).

Bacteria are unicellular structures, a few microns in length and enclosed in a membrane or cell wall. They exist in a variety of shapes and sizes: a bacterium or a bacillus is a straight rod; a vibrio is a curved rod (Fig. 43); a coccus is a spherical cell. The cell content, the protoplasm, performs many functions and activities which in multicellular animals (metazoa), mammals

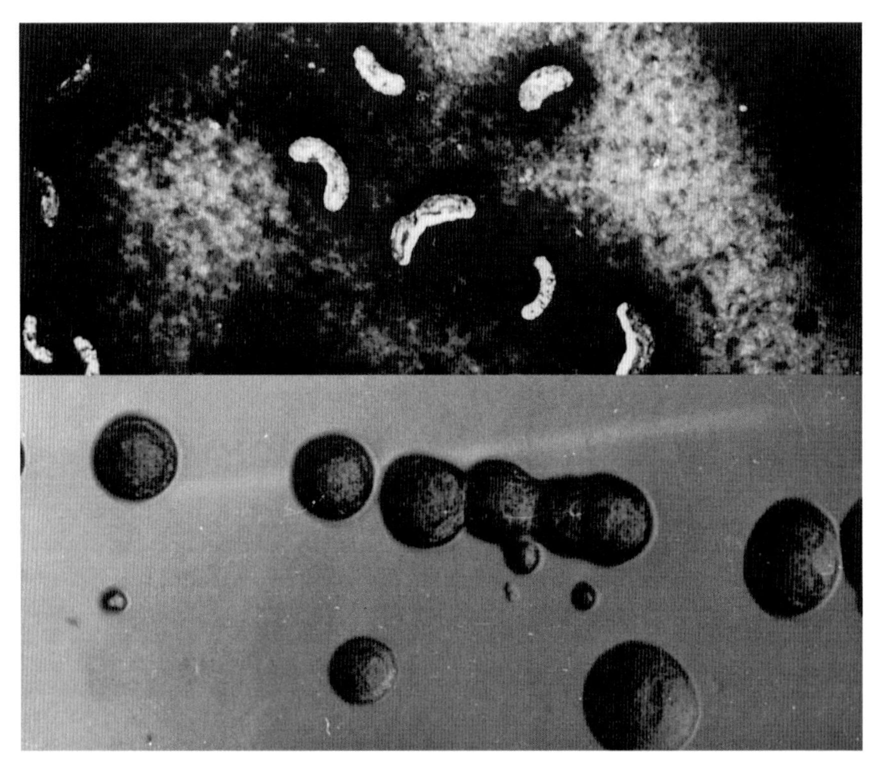

Fig. 43 *Above* Electron microscope photograph of cells of a tiny nitrifying bacterium isolated from a mountain soil (Snowdonia). This recently recognized *Nitrosovibrio* sp. is a small vibrio. (Phosphotungstate negative stain by N.W., magnification ca. 20,000). *Below* Photograph of colonies of a *Nitrobacter* sp., a nitrite-oxidizing bacterium isolated from an Argentinian soil. The circular colonies were grown on the surface of nitrite/agar medium (magnification x 8.)

for example, are carried out in specialized tissues or organs. A whole series of enzymes is contained within the bacterial cell. Enzymes operate in all the different biochemical activities of metabolism, digestion, respiration, excretion and growth. There is nuclear material, i.e. nucleic acids, but no organized cell nucleus. Nucleic acids are involved in the genetic processes of the cell and its reproduction. Bacteria multiply by simple fission; the cell grows, elongates and then divides into two. Under favourable conditions some species can divide within about twenty or thirty minutes. Weight for weight, bacterial tissue is metabolically much more active than that of higher animals, possibly because all the different activities are performed in the single cell. Moreover, because of the existence of the vast number of microbial species, it is not surprising that microorganisms as a whole are capable of a great range of biochemical activities. The current edition of Bergey's Manual of Determinative Bacteriology (1984) lists thousands of bacterial species, reflecting their astonishing diversity in form, nutritive and physiological capacities and ecological characteristics.

Many bacteria are surrounded by a capsule which serves as a protective coat and contains a kind of mucilage composed of different polysaccharides or other organic compounds. This mucilage, incidentally, causes organisms to adhere together as well as helping to cement soil particles into more stable soil crumbs, thus contributing to a useful soil structure. The capsular material is also an important factor which determines the specificity of the immunological properties of the microorganism.

The formation of spores or resting forms is another feature of many species of bacilli and bacteria. A spore is a resting stage which can survive adverse conditions, especially desiccation. When favourable growth conditions again prevail, then the spore can germinate and become a growing vegetative cell once more. Spores are usually spherical or ovoid in shape, often larger in diameter than the rod-like cell and so they assume a drumstick-like appearance under the microscope. *Bacillus* and *Clostridium* are the names of two genera of spore-forming bacteria the former being aerobic and the latter anaerobic. In some species the spore is seen as a swelling, sometimes in the middle and sometimes at one end of the cell. Fungi, actinomycetes and certain other organisms also form spores. As a rule, spores have a significantly lower water content than vegetative cells. Fungal and other spores, like the bacterial ones, germinate when conditions are favourable and resume active growth. All types of spores may be blown about by wind, along with dust and fine soil particles, and so can become distributed over distant soil areas and colonize them.

Water requirements of microorganisms vary; fungi and actinomycetes, for example, usually make do with less water than do most bacteria, and some can survive in quite dry environments. However, apart from spore formation, quite a number of bacterial species including non-spore formers, seem able to withstand desiccation for long periods, possibly because of adsorption on to the surfaces of clay minerals or other soil colloids. Certain autotrophic nitrifying bacteria, that are not spore formers, are known to remain viable in air-dried soil for long periods. One of us (N.W.) recovered nitrifying bacteria successfully from soil samples which, after drying at room temperatures, had been stored in sealed jars at Rothamsted Experimental Station for up to a century or more. When these soils were re-moistened, nitrification (i.e. oxidation of ammonia to nitrate) by the resuscitated organisms ensued normally and the bacteria in question could be isolated in pure culture and identified.

It is common practice, nowadays, to preserve reference cultures of microorganisms by a freeze-drying process. Organisms from pure cultures are suspended in suitable solutions, frozen quickly and then dried in vacuum from the frozen condition. Such dried cultures are stored at low temperatures, usually 2° to 4°C, when they remain viable and may be preserved for many years.

However, it is when we turn to the study of different aspects of microbial physiology that one can appreciate their complex characteristics and behaviour, and realize what they can do and how they affect their surroundings, in this case, the soil. Free oxygen is essential for the growth and respiration of many microorganisms which, accordingly, are termed 'aerobic'. There are many other species which can flourish only in the complete absence of gaseous oxygen. These are designated 'anaerobic organisms' or'anaerobes', and in their respiration and metabolism they make use of the chemically combined oxygen present in salts such as nitrates or sulphates. Many other

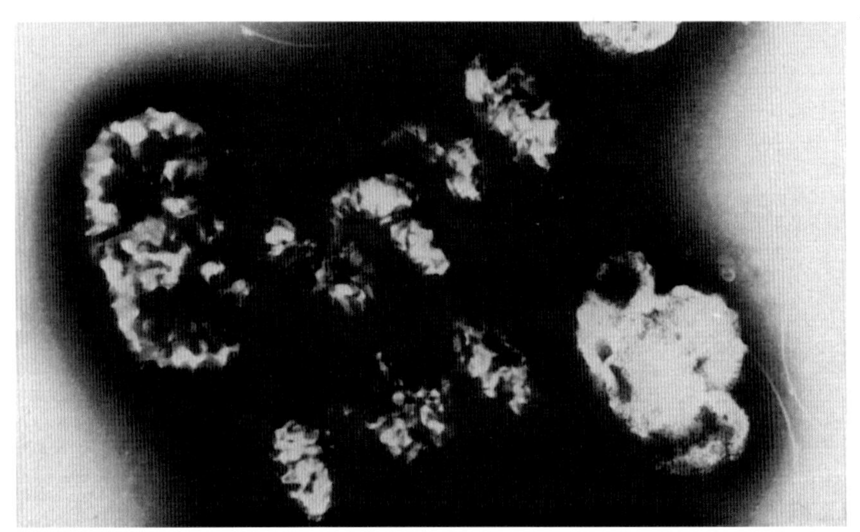

Fig. 44 Electron microscope photograph of cells of a *Nitrosospira* species, a nitrifying bacterium isolated from an acid woodland soil – Geescroft Wilderness, Rothamsted Experimental Station. Note the long coiled spring appearance. (Photograph N. W., magnification ca. 40,000.)

types of bacteria exist which, depending on conditions, can grow and function either with or without free oxygen, and these are referred to as facultative anaerobes. In their metabolism, microorganisms gain energy for growth and cell maintenance from the oxidation of different nutrients; a sugar, for example, under aerobic conditions is oxidized completely to carbon dioxide and water. On the other hand, under anaerobic conditions (i.e. absence of free oxygen) only a partial oxidation of the nutrient occurs forming carbon dioxide and water, and some incompletely oxidized substances remain in the form of intermediate oxidation products, such as alcohol and simple organic acids. The amount of energy thus mobilized is less than that theoretically possible, and therefore these anaerobic processes are distinctly less efficient than those in aerobic metabolism.

Temperature is also an important physical factor affecting the rate of microbial multiplication; different microbial species have different optimal growth temperatures. Some bacteria and a few fungi can grow at or near to freezing point and even exhibit their optimum growth at between 5° and 10°C; such species are termed psychrophilic and many of them are found in the sea. However, most bacterial species grow and flourish best at between 20° and 35°C and they are described as mesophilic organisms. Some microbes show optimum growth at between 45° and 50°C and there are other very specialized organisms which can grow even at 75° or 80°C. The latter are thermophilic organisms and they may be found, for example, in hot volcanic springs which often contain sulphur compounds and so represent a most extreme environment. Various bacteria and other microorganisms can tolerate and multiply in most unpromising conditions, for example in quite acid mine waters, and there are others which exist in alkaline environments, with or without free oxygen and often utilizing very unusual nutrients.

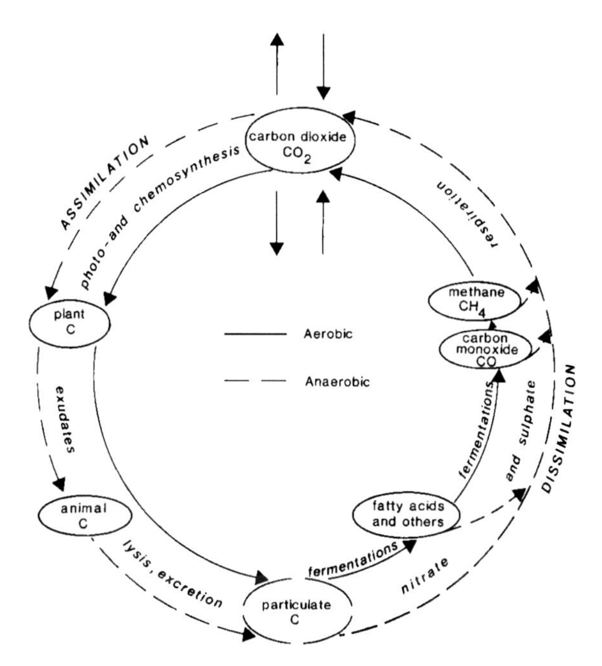

Fig. 45 Carbon cycle. Carbon dioxide is assimilated by green plants through photosynthesis and converted into organic substances. These are broken down (dissimilated) in the respiration and metabolism of plants, liberating carbon dioxide again as the final product. The presence of oxygen affects the course of C cycling, and its absence allows fermentations to proceed, forming gases such as methane (CH_4). (Adapted from G. Rheinheimer 1985.)

The reaction of a solution or a medium, that is to say whether it is acid, neutral or alkaline, is determined by the concentration of hydrogen ions (see chapter 1) and it is an important factor influencing microbial growth and the nature of microbial populations. As a rule, fungi seem to require more acid conditions (i.e. a lower pH value) than bacteria or actinomycetes which generally prefer neutral or slightly alkaline reactions (i.e. a pH range of, say, 6 to 8), but there are always exceptions (Fig. 44). Iron- and sulphur-oxidizing bacteria proliferate under quite acid conditions, where the pH value is around 1 or 2, as in acid mine waters or in the colliery wastes described in chapter 10. Differences in the reaction of soils are often reflected in their microbial populations.

Activities of soil microbes

Probably because of their minute size, simple unicellular structure and very varied metabolic activities, microorganisms can affect their surroundings in many ways; such effects in soils are of particular interest to us. An important example of these activities is the contribution they make, in association with the small soil fauna, to the breaking down of the mass of organic material which is constantly reaching the soil as plant litter and animal debris. The decomposition of organic matter in this way is an essential stage in the Carbon Cycle in nature; it is vital to the maintenance of life on this planet.

Plants and certain microorganisms produce organic substances from carbon dioxide with exploitation of solar energy in photosynthesis, or by chemosynthesis. Atmospheric carbon dioxide (the content of CO_2 in the atmosphere is about 0.03% by volume) is the carbon source for plants to synthesize their organic material. As much as 100 tonnes of organic dry matter per hectare may accumulate annually in tropical forests. Breakdown of this material and its eventual mineralization to CO_2 is a major process in the Carbon Cycle. Organic substances are the food for animals or microorganisms and, after their death, the organic matter in their bodies is broken down again, largely in the soil, with release of CO_2.

En route, some carbon compounds may be re-cycled by assimilation into the protoplasm of microorganisms and other organisms participating in the degradation of organic matter, but subsequently these organisms also die and are degraded. Thus the CO_2 of the atmosphere is derived partly from geological sources as in the combustion of coal or oil, and from biological sources, including the respiration of all living organisms. A simplified outline of the Carbon Cycle is presented in Figure 45.

Most water-soluble substances present in the soft, fleshy tissues of plants are decomposed fairly quickly, because many microorganisms can readily metabolize simple organic compounds, such as sugars. The more stable or polymerized plant structural components, for example, hemicelluloses, cellulose and partially lignified materials, are generally more resistant to degradation. Nevertheless, although cellulose is not a water-soluble substance, both it and hemicelluloses can be rapidly degraded by the hydrolytic action of enzymes called cellulases, which are produced and secreted by various soil bacteria and also small animals, such as snails. These enzymes break down cellulose or hemicellulose into their component sugars and, in turn, the latter are readily digested by other microorganisms.

Lignified or woody tissues are decomposed much more slowly and often it requires the combined activities of successions of different small soil animals and microorganisms to bring about their degradation. The decay and breakdown of tree stumps and dead branches, for example, takes a long time and is influenced by many factors, such as site and subsoil conditions, moisture content, temperature and the tree species. W. Kühnelt described wood degradation processes from the point of view of the various participating soil animals. Fungi are also involved in the successive stages of wood and lignin breakdown as described later.

What has been said so far serves to emphasize the extensive range of nutritive and metabolic activities possessed by microorganisms in general. For his part, man is interested usually in the effects of microorganisms on their surroundings and whether these effects are beneficial or whether they are harmful and result in undesirable spoilage. But from the viewpoint of the microorganisms, on the other hand, these decomposition activities may be regarded essentially as processes that provide them with nutrients. The nutritional requirements of microorganisms are very diverse.

Some bacteria, which are termed autotrophic, can satisfy their food requirements from a few simple inorganic compounds. They obtain carbon from carbon dioxide, nitrogen from compounds like ammonia or nitrates, whilst energy for growth and metabolism is acquired from the oxidation of substances such as hydrogen, sulphur, hydrogen sulphide, iron compounds,

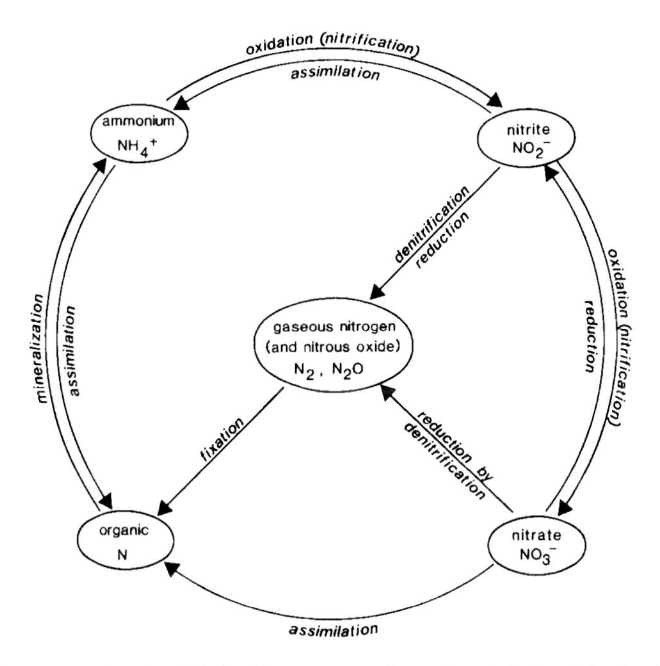

Fig. 46 Nitrogen cycle (simplified). Nitrogen gas from the air is fixed in the form of ammonia which is converted into organic compounds; the clockwise arrows indicate the breakdown of such compounds to ammonia (shown here as ammonium ions); this is oxidized by bacteria first to nitrite then to nitrate which is assimilated in plants to form organic nitrogen compounds again. Nitrate is reduced to nitrogen and nitrous oxide by other bacteria. Anti-clockwise arrows point to the reduction of nitrate to ammonia and, again, the assimilation of the latter into organic nitrogen compounds.

ammonia or nitrites. Although these bacteria do not use solar energy for photosynthesis, as when carbon dioxide is utilized by green plants, they are able to bring about the reduction of carbon dioxide by means of hydrogen produced within the cell, often from water molecules, in the course of their intermediary metabolism.

Perhaps the best known autotrophic soil bacteria are the ubiquitous nitrifying organisms. Of these, there are two distinct groups, represented by a few species belonging to four different genera. The first type includes those that oxidize ammonia to nitrite; the second type, namely the *Nitrobacter* species, oxidize nitrite to nitrate which is the final product of the nitrification process. Conversion of ammonia to nitrate is an important part of the Nitrogen Cycle in nature and Figure 46 depicts a simplified version. Nitrogen, the main gaseous component of the earth's atmosphere, can be combined or fixed by various microorganisms, initially forming ammonia which may then be incorporated into organic compounds in other organisms. The role of *Rhizobium* bacteria in root nodules was mentioned in chapter 3, but another important group are the blue-green Cyanobacteria. These may be present in rice paddy fields in vast numbers and may fix all the nitrogen required by the crop. This is an assimilation process. Complex organic nitrogen substances, present in plants, animals or other organisms, in turn can be broken down or mineralized by microorganisms into simpler substances including ammonia; the latter is then available to be converted into nitrate by nitrifying bacteria. Nitrates are taken up by plants and assimilated into complex nitrogen compounds again. Many microorganisms can also reduce nitrate to nitrite and then to ammonia; this is the reverse of the nitrification process. There are yet other microorganisms which decompose nitrate to form gaseous nitrogen, a process termed denitrification, to be described later, and so completing the cycle of nitrogen transformations.

Nitrifying bacteria are strictly aerobic organisms and require free gaseous oxygen for their respiration. They also need carbon dioxide to supply carbon and smaller amounts of a few simple salts to provide phosphorus, iron, sulphur and trace elements, e.g. perhaps copper. Nitrification occurs in nearly all soils. It proceeds best at a neutral reaction, although nitrifying bacteria grow quite slowly and reproduce only once or twice a day even under the optimum nutrient and temperature conditions possible in a laboratory. In soil, where conditions are far from optimal, growth is much slower. Ammonia that is released slowly *in situ* by the gradual breakdown of organic matter, is converted into nitrate, which is taken up by plants. The size of populations of nitrifying bacteria is largely determined by the quantity of available ammonia and the presence of oxygen and carbon dioxide. Consequently, nitrifying bacteria are generally to be found in the top two or three inches of soil. The application of ammonium sulphate to soil as a fertilizer may cause some acidification of the soil due to liberation of sulphuric acid and formation of nitric acid, but the extent of this depends on the buffering or ion exchange capacity of the soil as well as the presence of lime. Nitrification also takes place in the sea and other bodies of water, and especially in sewage and ammonia-containing effluents.

Nitrates, which are readily soluble in water, are not adsorbed on the soil clay minerals and so are easily washed out of soil by rain. Inorganic cations, including the ammonium cation (the form adopted by ammonia when dissolved in water) are adsorbed on clay minerals, and thus are not subject to leaching. To prevent nitrogen loss, it is desirable that nitrification should occur when plants are actively growing and able to take up nitrates as soon as they are formed (see chapter 9). Specific chemical inhibitors of nitrification (e.g. Didin, N-Serve) are now available commercially, and the practice of including such inhibitors in ammonium-containing fertilizers is gaining ground. It is now possible to apply such fertilizers in autumn or winter and for nitrification to be inhibited until, say, early spring when the sown crop begins to grow. Nitrification is also prevented for long periods by fumigation of the soil; this treatment may be performed in special cases, such as in tree nurseries or in citrus growing.

Another way in which nitrogen can be lost from soils is as a gas. The conversion of nitrate, via nitrite, to nitrogen gas is brought about by a variety of anaerobic bacteria by a reduction process known as 'denitrification', as briefly referred to above (see Nitrogen Cycle, Fig. 46). Sometimes very small amounts of the gas nitrous oxide (N_2O) and nitric oxide (NO) are also produced. Oxygen, like nitrogen, is very sparingly soluble in water, and so microbes which utilize dissolved oxygen have to rely on atmospheric oxygen being constantly dissolved in the soil water to replenish their supply. Anaerobic conditions may readily arise in soil due to the rapid consumption of

All microorganisms require moisture for normal growth and excessive dryness can seriously affect microbial populations. The spores or resting forms of fungi, actinomycetes and spore-forming bacteria and the cysts of amoeba, as mentioned earlier, can withstand dry conditions. Non-sporing bacteria can also sometimes survive dry conditions in soil. It is supposed that clay minerals or soil colloids can protect adsorbed microorganisms in some way. Nevertheless, generally speaking, when soils become very dry during hot weather many microorganisms and small delicate animals succumb and, in due course, their dead biomass becomes transformed into nutrients for other living organisms when favourable conditions again prevail.

Microorganisms can be killed in other ways than by desiccation; for instance, by heat treatment as in steam sterilization of soil, or by toxic chemicals, as in soil fumigation (see chapter 9). Provided that sterilization of the soil is not complete, the biomass of killed organisms will, in turn, be utilized by surviving microorganisms and eventually will be completely mineralized. This provides inorganic nutrients for crop plants and so there is some stimulation of plant growth, just as in the case of repeated drying and re-wetting of soil. Much of the carbon in the dead biomass is lost as carbon dioxide, but most of the nitrogen will remain in the ammonium form, which can be transformed into nitrate by nitrification. A small proportion of the nitrogen may remain in combined form in the soil humus or in soil organic matter. Trace elements and other minerals are mostly retained in the soil.

Fungi, yeasts, actinomycetes and algae

In addition to abundant bacterial species, soils contain many other microorganisms of which brief mention has already been made of fungi and actinomycetes. Microfungi grow as tubular filaments, called hyphae, and interwoven masses of hyphae are known as mycelium. Fungal mycelia are often visible to the naked eye. There are thousands of fungal species and some of them are typical soil dwellers. Usually fungi require preformed organic carbon compounds for their nutrition, so are heterotrophic organisms. They do not possess chlorophyll and therefore are not photosynthetic organisms. Although often very diverse, their nutritive requirements are generally fairly simple. In morphology and mode of reproduction fungi show great variety. They reproduce by producing spores, some of which are quite small, a few microns in diameter, but others, like the chlamydospores and sclerotia can be much larger, up to 100 microns in size. Spores are formed both asexually and, in many instances, by a sexual process.

Four major groups of fungi may be distinguished: phycomycetes, ascomycetes, basidiomycetes and fungi imperfecti. Phycomycetes are typical moulds. They often grow quite rapidly and form spores in large numbers. In the course of sexual reproduction, short lateral hyphae come together and fuse forming zygospores. Phycomycetes commonly found in soils include several *Mucor* and *Mortierella* species; some aquatic species also may occur in water-logged soils.

Typical ascomycetes produce, in sexual reproduction, eight spores that are contained in a tubular or a spherical structure, known as an ascus. Asci may be grouped together either in a disc-like structure or in a flask-shaped organ, called a perithecium. Ascomycetes also produce vegetative type spores. *Chaetomium* is a typical soil ascomycete genus and it is an active cellulose decomposer. Frequently, many soil ascomycetes are associated with dung or animal droppings.

The formation of a club-shaped hypha, called a basidium which bears four spores on the end of short stalks, is a characteristic of the basidiomycetes. Basidia may be grouped on the surfaces of gills of a stalked cap, as in the toadstools, or in a spherical body, as in the puffballs. Soil basidiomycetes are not well characterized; although possibly hundreds of species may grow in soil or on plant debris their presence is only revealed when they form fruiting bodies. There are two main divisions, the gasteromycetes which have their fruiting layers concealed internally, and the hymenomycetes, in which the fruiting layers are open to the air. The study and determination of the many species is a matter for the specialist. *Rhizoctonia* species are common in soils. Forest soils harbour many hymenomycetes (i.e. the mushroom type fungi) and gasteromycetes (i.e the puffball type). In addition, mycorrhiza on tree roots are very widespread as described in chapter 3.

Marasmius oreades is an example of a fungus which causes 'fairy rings' on lawns or pastures. The fungal hyphae grow out radially from a point where a colony of the fungus has become established and, eventually, they become manifest in a ring of deeper coloured grass. As the years go by, this circle of mycelium grows steadily wider, sometimes at a rate of several inches per year. Several other species of fungi also develop rings of growth in this way. It is possible that the darker green colour of the grass is caused by some enrichment of the soil in nitrogen compounds where fungal growth has occurred. A bare circular area in the soil surface, bordering the darker green grass can be the result of the grass roots being covered by the fungal mycelia. Basidiomycetes frequently produce mycelia which can remain viable for long periods and so, when they have colonized some substrate or other, may still continue growing towards other substrates some distance away. Fairy ring type basidiomycetes tend to migrate in pasture soil in this manner (Figure 48).

Some fungal species that belong to the family Zoopagaceae are predators of nematodes, for example *Arthrobotrys oligospora*; there are others which feed on amoebae. Trapping of nematode worms in a ring of fungal hyphae is quite ingenious. Some animal-trapping fungi possess interwoven hyphae to which the nematodes adhere; the fungus then dissolves the nematode cuticle, drives hyphae into the nematode's body and absorbs the internal organs. In the fungus *Dactylaria brochophaga*, collar-like outgrowths are produced out of three cells of the fungal hypha and develop into a constricting ring trap. When a nematode passes its head into the opening of the ring, the cells are stimulated and swell rapidly thus tightly gripping the nematode. The action is very quick: the nematode is trapped, dies, and its body contents are absorbed by the fungus.

Yeasts are generally considered as fungi and they occur frequently in soil. They are oval or spherical unicellular organisms, mostly about 5 to 10 microns in diameter. Usually they reproduce by a budding process; a small outgrowth develops on the periphery of the cell, it continues to grow until it is large enough to separate from the mother cell and then becomes a new, independent yeast organism. Many yeasts also reproduce by a sexual process. Some yeasts grow filamentous hyphae and produce mycelia, just as do moulds. There are also yeasts, for example *Rhodotorula* species, which are

Fig. 48 'Fairy ring' of toadstools. The photograph shows the fruiting bodies of three species of mycorrhizal fungi on roots of birch. The main, outer, ring is formed by *Hebeloma crustuliniforme*; four large whitish toadstools, *Lactarius pubescens* form a partial inner ring, while a single brownish toadstool nearest the tree is *Leccinum roseofracta*. (Photograph P.A.Mason.)

capable of fixing atmospheric nitrogen. Although, in general, yeast and fungi flourish best under aerobic conditions, fermentations by yeasts proceed anaerobically. Bakers' yeast and beer yeasts are different races or varieties of the species *Saccharomyces cerevisiae*. Both fungi and yeasts can usually tolerate a fairly wide range of pH values.

From the ecological point of view, fungi, as in the case of bacteria, can be classified on the basis of their food requirements: for example, 'sugar fungi' require only sugars or simple carbon compounds. Other types of fungi can assimilate or decompose cellulose or lignin. Basidiomycetes are typical lignin-decomposing fungi, although they grow on wood or lignin materials only slowly and often can assimilate lignin better when it is associated with some more available nutrient, such as cellulose.

Two main groups of fungal species, known as 'white rots' and 'brown rots', are recognized in the decay and rotting of wooden structures, especially in wet situations. Some typical microscopic moulds are wood-attacking fungi and others, like the bracket fungi, may form large macroscopic growths on the trunks or branches of trees. Lignin is the characteristic chemical component of woody tissues, and is manufactured in plants by the enzymatic oxidation and polymerization of phenylpropane derivatives. It is understandable that stable, complex, insoluble substances of this kind are not easily attacked and decomposed.

Apart from fungi, few microorganisms can digest wood or lignin directly. Some termite species degrade wood and lignified tissues with the assistance of certain fungal species which they cultivate in a kind of fungal garden within the termite mound. Other insoluble materials, for example, chitin which is present in the cuticle and exoskeletons of insects, woodlice and other arthropods, are broken down only slowly. Waxes also are degraded only slowly.

Fungi imperfecti are so-called because no sexual stages have been recognized in them, and their precise classification, accordingly, is rather problematical. They appear to reproduce only asexually. Many soil fungi belong to this group which includes, for example, typical moulds like *Penicillium* and *Aspergillus* species. The latter species of moulds can grow on most available substrates in soil, such as pieces of plant litter or other organic debris and they always produce multitudes of spores. *Mucor* species also colonize litter and similar material, form many spores, and from there they often grow further into surrounding areas.

Actinomycetes, which are a large group of filamentous organisms, are somewhat intermediate between bacteria and fungi. As individual cells, they are similar in size to common bacteria but they can grow into long branching filaments and produce a ramifying network resembling fungal mycelia. The actinomycete mycelium may also later fragment into smaller elements like bacteria in size and appearance. The hyphal filaments are usually about one micron in diameter and may sometimes grow into aerial filaments. Chains of spores are formed simply by the growth of cross walls. Actinomycetes are variable in morphology and other characteristics so their identification and classification is often difficult. Some species give off an earthy odour due to chemical compounds known as geosmins; the smell of newly dug soil, most likely, is attributable to this. Actinomycetes have assumed great importance in recent years because of their propensity for producing antibiotic substances. It is particularly interesting that many antibiotics derived from actinomycetes or streptomycetes possess unusual and complicated chemical constitutions.

S.A.Waksman and his colleagues discovered the first such antibiotic, streptomycin, which had a significant effect against the tubercle bacillus (the causative agent of tuberculosis). Since then much research has been devoted to the isolation and investigation of numerous soil actinomycetes. Many such antibiotics have now been applied in human and veterinary medecine, or in agriculture. Attempts to demonstrate the presence or production of antibiotics in soil which may possibly be responsible for antagonism between different microorganisms have met with little success. One example is griseofulvin, but the amounts found are too small to have much importance.

Among the actinomycetesa, several subgroups are known and these are classified generally on the basis of spore and spore formation characteristics. Most soil species belong to the genera *Nocardia*, *Streptomyces* and *Micromonospora*. Species of the latter genus are widespread in soil and many are active cellulose decomposers. *Nocardia* species are especially versatile in their metabolism and can degrade many different carbon compounds, including hydrocarbons, tannins, lignin, rubber and several synthetic plant protection chemicals. Chapter 3 described the nitrogen fixing ability of *Frankia* in association with plant roots.

Myxobacteria represent another group of peculiar soil microorganisms with a complicated life cycle. They consume other typical living soil bacteria and, therefore, can be regarded as predators. They are found and can be isolated from sheep or rabbit dung pellets. Myxobacteria can be grown on plate surface cultures of various ordinary bacteria. In their life cycle, masses of myxobacterial cells come together in swarms and gradually develop into fruiting bodies, which are large enough to be seen with the naked eye and are often brightly coloured. Fruiting body formation seems to occur in response to declining nutrient supply. The swarms of vegetative cells, which possess gliding motility, gradually differentiate into a fruiting body with a stalk and a head. Most of the cells accumulate in the fruiting body head where they become transformed into myxospores. In due course, germination of the myxospore signals a fresh round of the life cycle.

Algae, which include both unicellular and filamentous species, are often found in soil. In common with green plants, they are photosynthetic organisms which contain chlorophyll and assimilate carbon dioxide in the presence of light. Yellow-green forms belonging to the class Xanthophyceae are one of the most common groups. Because of their dependence on light, they grow on or near the soil surface. The blue-green Cyanobacteria were formerly considered to be algae; their importance in rice paddy fields has already been mentioned.

Under suitable conditions, extensive blooms of coloured algae may develop in summer on the surface of waters, both inland and marine. Algal cells often contain other pigments besides chlorophyll so that deep red coloured algal blooms may form on the surface of waters, as for example, on Lake Tovel in the Dolomite area of Italy.

This brief account gives only a general impression of the better known members of the very diverse microbial world and its wide range of activities. In spite of their small size, microorganisms are metabolically very active and consequently play an important role in the living soil. Their rapid rate of multiplication enables many generations to be produced in a short time span and also for mutations and genetic changes to occur with some frequency. Adaptations and metabolic changes result from these genetic alterations. Great advances have been made in recent years in microbial genetics and the transfer of genes from one bacterial species to another has become possible. Such transfers can also be mediated by viable fractions of the cell known as plasmids which are motile and can be incorporated in other cells. Bacteria or other microorganisms thus acquire new enzymes.

Various enzymes can often be found in soil; they may be secreted directly from microorganisms, for example cellulases, or they are liberated on the disintegration of some bacteria or other. The enzyme urease, which catalyses the conversion of urea into ammonium carbonate, may often be detected in soil. It is essential to the utilization of urea fertilizers. Bacterial enzymes are also important in the breakdown of pesticides in the soil (see chapter 9).

7

Natural Habitats

A framework of classification for British soils was outlined in chapter 2. This scheme is based on the idea that one can recognize links and relationships between different soil classes resulting from the interaction of consistent soil-forming factors. Mature profiles, typical of these different soil classes, only develop when natural trends operate over a sufficient length of time without significant human interference. Completely natural situations that provide these conditions are rare in Britain, especially in lowland regions. Most uncultivated sites, even in what appear to be natural habitats, are more correctly thought of as semi-natural because of their past cultivation or other land use influences. Even such semi-natural soils are quite uncommon in the parts of Britain dominated by intensive agriculture. Everywhere, however, they are important 'benchmark' soils which justify protection, as we said in chapter 2, as they are part of the natural environment that cannot be re-created. By displaying the least modified expression of natural soil development appropriate to a particular site, these type examples give a foundation for soil classification. They also help us to understand the processes of soil development and the modifications produced by cultivation.

Contrasts between soil profiles of cultivated and uncultivated variants of the same class are most conspicuous in their upper organic horizons. Surface horizons are, of course, those most disturbed by cultivations, and they are also the hub of biological activity. Their character is produced by the differing trends and rates at which soil organisms return plant litter to the nutrient and energy bank of the soil. The way these trends are expressed is largely determined by the physical and chemical properties of the underlying mineral soil, but it is also affected by the vegetation cover. Natural or imposed vegetation changes influence soil development directly, and also indirectly through the soil fauna, as illustrated below. This is a two-way interaction, with soil characteristics controlling, and being the consequence of, biological activities. In some soil classes, e.g. Brown Earths, organic matter becomes intimately mixed with the mineral soil as described in chapter 2, while in others, e.g. Podzols, organic horizons remain sharply distinct.

This interacting 'world of the soil' has in the past been approached more often as if it were two parallel but separate worlds rather than as a single entity; usually, the pedological and biological approaches have made only glancing contact with each other. One can readily understand this situation since specialists in these two fields have generally come from quite different scientific backgrounds. There are also intrinsic difficulties in relating the scale of biological activities to soil properties. To encompass the activities of microorganisms, mites and moles, one must range from mineral and organic particle surfaces, through pore spaces of increasing size, to the large-scale physical character of soils. This ideal multi-level view of the soil is,

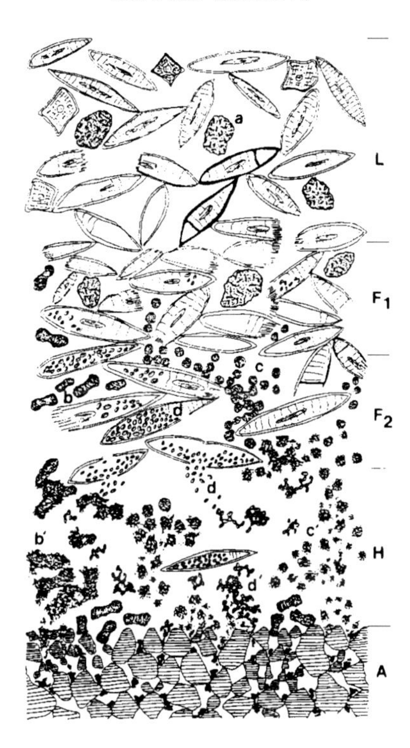

Fig. 49 Development of moder humus from Douglas fir needle litter on a sandy soil, showing litter layer L, decomposition or fermentation layer F1, F2, and humified layer H above the mineral A horizon. Fresh and aging excrements of: a, a' = leatherjackets (crane fly larvae, Tipulidae), b, b' = litter feeding *Adela* caterpillars, c, c' = fungus gnat larvae (Mycetophilidae), d, d' = beetle mite larvae (Phthiracaridae). (From L. Bal 1970.)

unfortunately, hardly ever practicable but it remains a desirable objective. There is certainly still much scope for studies of the relationships between species and their soil profile environment. Meanwhile, this chapter considers some work on the interactions between soil animals, soil profiles and vegetation, in woodland, moorland and grassland habitats.

Woodlands

The biological importance of litter decay is most conspicuous in woodland. In a detailed study in Holland, L. Bal looked at the sequence of organic horizons formed as fresh plant litter was progressively attacked by soil organisms. He compared the moder humus profiles developed under red oak *Quercus borealis* and Douglas fir *Pseudotsuga menziesii* when planted on very acid sandy soils. Under both species, the depth of humus was similar – about six centimetres – but the profiles differed in detail. In the fir profile (Fig. 49), there was an approximately equal division into three zones. The surface litter, or L layer, was composed of only slightly modified fir needles. This merged into an F layer in which the needles were susceptible to attack by soil animals, and an increasing complement of faunal droppings appeared. The third zone

Fig. 50 Leaf discs eaten by various invertebrates. The discs had been cut from oak and beech, placed in nylon mesh bags of 7 mm, 1 mm and 0.5 mm openings and buried in the soil. (From C. A. Edwards & G. W. Heath 1963.)

was a fully humified H layer, dominated by weathered droppings that had passed through more than one cycle of ingestion and excretion, with only skeletal fragments of the initial needle litter remaining. The transition between this humus and the underlying mineral horizon was fairly sharp, there being only minor physical inwash or faunal transport of the humus particles among the sand grains.

In contrast, the more rapidily decomposed oak litter produced a rather thinner L layer. This passed again into an F layer of partially decomposed and compacted leaves mingled with faunal droppings but, below this, the completely humified H horizon occupied about half the thickness of the whole humus profile, and was more conspicuously mixed with sand grains from the mineral horizon.

In looking at the fauna inhabiting these humus profiles, Bal distinguished characteristic groups of organisms that had a dominant influence in creating the different types of moder. Fly larvae, mites and enchytraeid worms were common to both, though the population size and species differed. Both the oak and conifer litter had first to be rendered palatable by microbial attack but the fir needles needed a longer period before they were acceptable.

A similar trend of humus profile development is characteristic of Brown Podzolic Soils in Britain, under, for example, birch and oak woodlands in the uplands, or some woodlands and heaths on acidic soils in the lowlands. On the less acid and more fertile soils of lowland Britain, woodland litter is incorporated faster and more thoroughly. This results in the mull humus forms characteristic of Brown Earths and Calcareous Soils. The rapid recycling of nutrients that is a feature of mull humus soils is important in maintaining natural fertility. In his work on the biology of woodland soils, K.L.Bocock pointed out that 75 percent of the annual nutrient requirements of trees in closed canopy woodlands could be met by the decomposition of one year's litter fall. His colleague J.Satchell estimated that the annual nitrogen turnover due to earthworms in a fertile mull soil under mixed woodland was four times greater than the nitrogen in the annual litter fall.

Other early work by the Nature Conservancy in the Lake District compared the respective attack by organisms on freshly fallen oak and ash litter placed on the soil surface in nylon mesh bags. One site was mixed deciduous coppice (mainly ash and hazel) on a mull humus over limestone; the other was oak-birch coppice on a more acid moder over slaty rocks. At the mull site, ash leaflets had been skeletonized within the first six months, and some had even been dragged from the nets by large earthworm species. On the moder humus site, the leaves remained in the nets and were extensively attacked by mites, springtails, enchytraeids and small earthworms (Fig. 50). Oak leaf litter showed similar contrasts between sites, but with a delay before substantial faunal attack occurred. Again, some initial weathering and microbial attack was needed to remove some of the tannins and render these palatable.

Turning to a Northamptonshire site, Bedford Purlieus near Peterborough is a complex woodland with a mixture of coppice regrowth, planted deciduous and conifer species, and naturally regenerated trees. The wood is notable not only for its diverse flora and fauna but also for providing important benchmark examples of undisturbed lowland soils which overlie strongly contrasting parent materials. At one end of the soil chemical range, calcareous Rendzinas are disturbed by high populations of earthworms and moles. The worms incorporate the humus deeply into the profile and contribute to a strong granular crumb structure, while the moles continually bring limestone fragments up from below.

The adjacent soils over acidic sands and silts include Brown Earths and poorly drained Gleys. The larger burrowing and surface-casting earthworms are absent, and differences in the humus profiles are related to the plant cover. Sweet chestnut *Castanea sativa* and dense bracken *Pteridium aquilinum* give moder humus with deep L layers. Where bracken is sparser, and particularly where birch is present, mull phases occur. Here again, the nutritional quality of the leaf litter affects the earthworm fauna which plays such a crucial role in the incorporation of organic material into the profile. A similar picture is seen in some moorland and upland habitats.

Moorland and hill

J.Miles and colleagues studied the soil contrasts associated with heather moor and patches of birch woodland in Scotland. This mosaic is a widespread feature of the central and eastern Cairngorms. In Miles' study, heather-dominated areas were mainly on Podzol soils with mor humus. Where birch trees *Betula pendula* and *B. pubescens* colonize an area, the heather *Calluna vulgaris* gradually declines and is replaced by grasses and herbs typical of open woodland. These vegetation changes are accompanied by progressive changes in soil conditions which appear to be due to the relatively nutrient-rich litter of the birch and its associated ground flora compared to that of the heather. The mor humus alters towards a mull-like form, the podzolic leached mineral horizon becomes obscured, the soil acidity is reduced, and there is a substantial increase in earthworm numbers.

Soil 'improvement', in the sense of greater decomposer activity and hence more rapid nutrient cycling, does not progress continually in the Scottish situation to maintain a woodland community at a particular location. There is a natural cycle determined by a decline in the vigour of birch, and its subsequent death at 70–90 years old. Birch does not establish itself successfully under a closed canopy with the result that, within the established woodland, there are no younger generations of birch to maintain the wood. As the old birch trees die, heather spreads back from the adjacent moor. Meantime, the

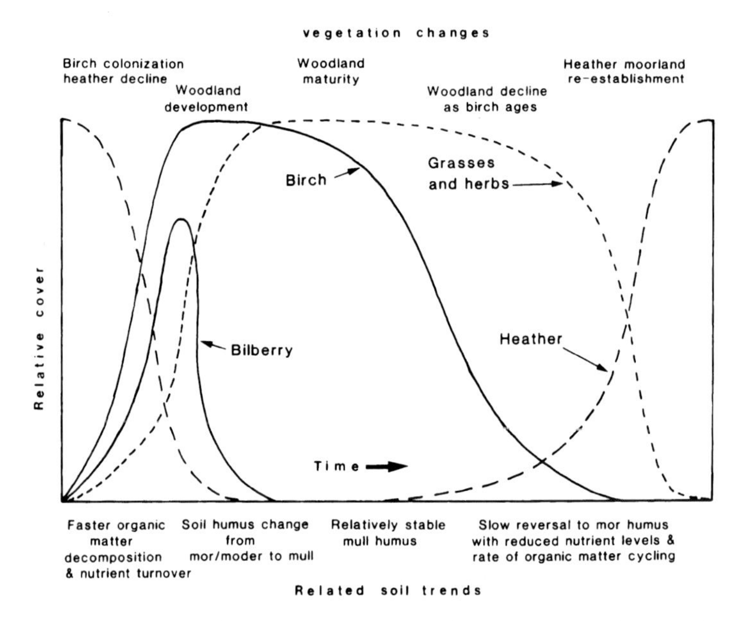

Fig. 51 Changes in tree and ground vegetation during growth and decline phases of birch woodland in Scottish heather moorland, and their effect on soil development. (Adapted from J. Miles 1981.)

birch will have colonized another area and a new young wood will be growing up. Figure 51 illustrates this cycle of vegetation and the related soil changes.

On a typical Morayshire site, the increase in earthworms during this cycle was from about one per square metre under heather, to five per square metre under 18 year old birch, and a peak of 127 per square metre under 38 year old birch. After this, the numbers of worms declined again as the birch aged and began to die. The earthworm species were *Lumbricus rubellus* initially but *L. terrestris* and *Aporrectodea* species appeared as the exchangeable calcium level increased in the upper soil horizons. This rise in calcium levels may have resulted from uptake by birch roots reaching lower soil depths than those tapped by the heather, but perhaps also from increased weathering of calcium-containing minerals *in situ*.

Elsewhere in Britain, there are moorlands where this natural cycle does not occur. On the North Yorks Moors, the idea of birch as a soil improver had been put forward by G.W.Dimbleby in 1952, and a long-term experiment with planted birch was set up on the Peaty Podzol soil over acid sandstone. However, the podzolic mineral horizons remained unchanged even after 30 years; mor surface humus, which had been destroyed to make a seed bed for the birch, had re-formed; the pH remained very low, and no earthworm species associated with mull humus had appeared. In these circumstances the birch grew very poorly, and the ground flora remained dominated by acidic moorland species such as wavy hair grass *Deschampsia flexuosa* and bilberry *Vaccinium myrtilus*. This site was probably too deficient in calcium for most earthworms, unlike the Scottish situation where calcium was evidently sufficient to support low densities in the open moorland, and worms were thus able to spread and multiply as the birch litter increased.

In upland areas where less wet and cold climates have, with the aid of subsidies, justified agricultural improvement, 'reclamation' of moorland to grassland has been a prominent land use change. On Exmoor, in southwest England, E. Maltby followed the effects on soil microbial populations that resulted from such reclamation of Peaty Gley and Peaty Podzol soils. Cultivation and fertilizer applications caused an immediate upsurge in bacterial numbers of about a hundredfold as compared with those in unchanged moorland soils. In the early years after reclamation, bacteria remained dominant among microorganisms, but actinomycetes increased relatively in soils reclaimed for longer periods. They seemed to react more slowly to changed environments. The physical, chemical and biological effects of conversion to grassland produced mull humus surface horizons like those seen in typical Brown Earths on similar parent materials at lower altitudes. Beneath the zone modified by cultivation, the lower horizons remained as relic features persisting from the initial Podzol and Gley profiles.

If intensive management ceases in such sites, so that lime is no longer being applied, the vegetation gradually reverts towards moorland grass and heath communities. The acid mor humus of these characteristic moorland soils may then gradually redevelop. Biologically, the relative numbers of bacteria and actinomycetes can be used to assess the degree to which reversion has proceeded from intensively managed grassland towards moor. Chapter 10 describes the way the balance of microorganisms is likewise affected during land reclamation.

Moving to hill areas with higher rainfall, we can look at some contrasts on the Snowdon National Nature Reserve in North Wales. As described earlier from Bedford Purlieus, there are sharply different parent materials here which are reflected in very distinct soils.

At one geological extreme, hard, acidic volcanic lavas and ashes (rhyolites), have remained as outcrops virtually unweathered in the 10,000 or so years since ice finally retreated from its last strongholds in the region. In the montane climate of high rainfall and low temperatures, shallow Peat Ranker soils have formed on exposed surfaces by the accumulation of organic matter. Initially, lichens and small mosses colonize depressions in the rock surface, but, as organic matter builds up, species such as heath rush *Juncus squarrosus* and *Sphagnum* mosses become dominant. These soils typically have extremely low pH values between 3.8 and 4.2.

The erosive power of glacier ice ground up and transported a mixture of these rhyolite rocks with other rock types, depositing the material as glacial drifts in valley floor situations within the Snowdon reserve. A complex of different soil classes now occupies the crests and slopes of the hummocky drift terrain, and the hollows between these hummocks. This association of Peaty Podzols on crests and upper slopes, Peaty Gleys on lower slopes, and deep peat in the hollows, supports different vegetation communities, controlled by variations in soil drainage. All these soils are slightly less acid than the shallow peat soils, with pH values generally between 4.1 and 4.7, but all have peaty surface horizons. At the drier end, the peaty podzols carry acid grassland dominated by mat-grass *Nardus stricta* and sheep's fescue *Festuca ovina*. Rushes, sedges and mosses are prominent in the peaty gley soils, and the deep hill peats are dominated by cotton-grass *Eriophorum* and *Sphagnum* mosses.

In contrast to the virtually inert rhyolites, other outcropping rock types on Snowdon weather chemically, or break down physically, very fast. Some slightly calcareous soft volcanic ashes break up within tens of years when freshly exposed. These rocks contain minerals which, on weathering, release calcium and magnesium. As a result, Brown Earth soils with mull humus (Fig. 13, p. 40) occur on sloping sites side by side with the peaty moorland soils on the more acid rocks and drifts. This is unusual as Brown Earths are not the normal soil response to such cold wet climates; the rainfall here is about 3,300 millimetres a year (130 inches). They support a relatively productive hill grassland with sheep's fescue and bent grass *Agrostis tenuis* accompanied by clover and a variety of herbaceous species. The rock type and sloping ground initially produce soils that are only moderately acid (pH 5.2–5.8) and are freely drained, but their profile character is maintained by physical and biological cycling of material.

As with the woodland soils on limestone at Bedford Purlieus, moles and earthworms play a key role in this process by bringing mineral material to the soil surface. Their activity counteracts the leaching and loss of soil calcium, and the consequent soil acidification, which the high rainfall and free drainage would otherwise produce. Additions of freshly weathered material from rock exposures at the head of the slope continually replenish the upper part of the soil profile. The soil depth remains largely unchanged, however, since fine particles are washed right down the slope to accumulate ultimately as thick silty sediments on the lake floor at the foot. The sheep that graze this area also help to maintain the level of plant nutrients in the soil by their droppings. They can select where they feed on these open hill grazings, and naturally favour the more palatable *Agrostis-Festuca* grassland to the less nutritious, more widespread plant communities. In this example, grazing livestock, moles and soil macro-fauna all influence the soil condition as well as react to it.

Mole activity was considered in chapter 5 but, in concluding this look at the biological features of the soil on moors and hills, we can summarize a regional study of mole distribution on open hill grazing in northern Snowdonia. As we said earlier, the frequency of mole-hills is not a sure guide to mole numbers. However, in this upland environment which is generally low in nutrients, mole-hills are a conspicuous indicator of soils that are fertile enough to support earthworms and therefore a grassland community more favoured by sheep. In the Snowdonia study, mole-hills were recorded (as either present or absent) in an extensive series of contrasting areas established for a sheep grazing census. Out of 48 study areas, 19 had mole-hills, and, in these, the soils were all Brown Earths or Brown Podzolic Soils. They had surface horizons of mull humus or were intermediate between mull and moder, with a mean pH of 4.8. Most sites without evidence of moles had soils with peaty surface horizons and were significantly more acid, with pH around 4.3. A few sites were suitable for worms but could not support moles because their very stony soils prevented tunnel construction.

Grassland habitats

Within these generally infertile moorland and hill habitats, the grasslands

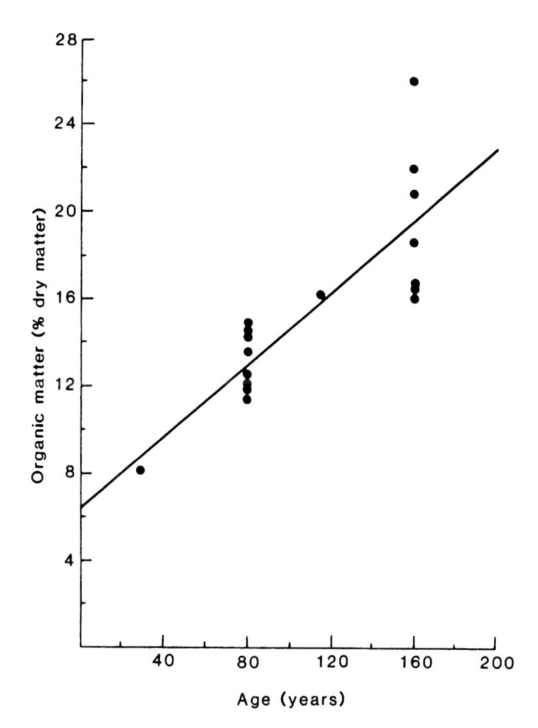

Fig. 52 Increase in soil organic matter content in chalk grassland over a period of 130 years at Porton Down, Wiltshire. (From T.C.E. Wells, J. Sheail, D.F. Ball & L.K. Ward 1976.)

that occur on Brown Earth soils over base-rich rocks can be considered as islands. But even in the lowlands, semi-natural grasslands have become scarce features. The limited areas that remain are virtually all within conserved sites. Communities of native species which simulate old grassland may be re-established on land that is 'set aside' under present agricultural policies. However, the creation of natural grassland soils would need many times the five year period at present allotted to the scheme, as the next example shows.

One might expect that lowland grasslands would be most extensive on the chalk downlands of southern England, but, even here, it is rare to find sites that have not been ploughed and used for cereals in recent years. An important and virtually unique area has been protected from this land use change by its inclusion within the Porton Ranges of the Ministry of Defence in Wiltshire. Here, a multi-disciplinary project was carried out by T.C.E.Wells and colleagues in the Nature Conservancy (later of the Institute of Terrestrial Ecology). This study, already touched on in the section on ants (chapter 4), explored the relationships between land use history, vegetation, soil fauna, and soil properties.

Cultivation last occurred on different parts of the area at different periods. The land had been in Government hands since 1916, and this date marked the end of any substantial agricultural use, though range activities caused

Fig. 53 Ant-hills of *Lasius flavus* on chalk grassland at Porton Down in Wiltshire, covered with rock-rose. (See Table 5, p. 86; Photograph B.N.K.D.)

local ground disturbance in some parts. The most recent, and very limited, cultivation had taken place about 30 years before the study was made in the early 1970s. The longest established grassland had not been cultivated for about 150 years. The historical evidence showed that cropping continued at different locations until the soil nutrient status and its organic matter content had declined to levels at which, in the conditions of the time, further cropping was uneconomic. The land was then left to revert to grass naturally. At most sites, the abandoned crop land passed through a succession of grassland types during which the soils regained their organic matter and nutrient content from the accumulation and incorporation of grass litter. The annual production could be fully returned to the soil since there was virtually no domestic livestock to remove it.

The organic matter content of soil is readily measured, and so it was possible to relate its gradual increase at Porton to the time since cultivation ceased. Figure 52 shows this increase in the top ten centimetres of soil, from samples taken along seventeen transects through different vegetation types of known age. From an extrapolated starting point of about six percent, the organic matter reached about eight percent after 30 years, and from 16 to 26 per cent on areas that had reverted to grassland 150 years ago. This gave an average annual increment of 0.08 percent organic matter over the whole period.

As mentioned earlier, the yellow ant *Lasius flavus* is abundant in these uncultivated grasslands. It makes an important contribution to the cycling of organic matter and nutrients, and also has significant effects upon the soil profile and vegetation through their mound-building activities (Fig. 53). The scale of these effects can be illustrated by estimating the depth to which the ground surface would be covered by the soil in the ant-hills if the mounds were evenly spread. This depth reaches a maximum of 20 millimetres on the oldest undisturbed grassland, but this underestimates the overall effect since it ignores reworked soil below the present ground surface, and also the past product of abandoned and decayed mounds. However, it shows the scale of physical modification achieved by a small creature when present in large numbers.

To summarize the picture at Porton, the present substantial variation in chemistry within soils of a single class is a direct consequence of ecological succession and biological activity acting over varying lengths of time. It is not the result of distinct parent materials, and hence different soil classes, within a site.

In the United States, the entire A horizon of some virgin Brown Podzolic Soils consists of material brought to the surface from the B horizon by ants. This material reaches thicknesses of 30–40 centimetres which have accumulated at an estimated rate of one centimetre per century. It is thus not only the larger earthworms that can stratify soils and develop clear profile characteristics. Of course, if we were not concentrating on British habitats, then the effects of termites are even more conspicuous. They can build mounds up to 2–3 metres high and 6–10 metres across in long-established locations.

Our consideration of soils and the role of organisms in semi-natural soils began with a look at the detail of humus profile formation in a woodland situation. Similarly, for grasslands, earlier work by B. Barratt produced a 'classification of humus forms' for grassland soils from a wide range of English and New Zealand sites. She used thin sections to identify and illustrate the relative frequency of fungal hyphae and resting bodies, faunal droppings and partially eaten leaf fragments and roots within the different humus forms. In most grassland soils with mull humus and high base status (pH above 5.5), there is no conspicuous and discrete humus layer. Instead, a thin, rapidly disappearing litter layer passes into a mineral horizon intensively worked by earthworms. On somewhat more acid Brown Earths (pH 4.5–5.0), an H layer is present with enchytraeid worms more prominent in the soil fauna. Moder humus profiles in grasslands that are intermediate in profile character between Brown Earths and Brown Podzoloic Soils typically have distinct L, F and H layers as in the Dutch woodland soils. The well humified H horizon is due to microbiological activity and the action of mites and enchytraeids, rather than earthworms. At the acidic end of this spectrum of humus types on grassy moorland, the surface horizons are mor humus, with mites and springtails conspicuous in the soil fauna.

Barratt also briefly considered some humus forms of poorly drained grassland soils. Her samples were also principally variants of mor, but in general little seems to have been written about the soil biology of natural wetland habitats in Britain.

Finally, we can look at how systematic variation in soil moisture conditions is reflected in grassland vegetation. This effect is clearly seen in old meadows that show a wave-like sculpturing known as ridge-and-furrow. Winter snow lying in the furrows displays the patterned relief very clearly (Fig. 54). This agricultural system was laid out originally for tillage, partly as a means of apportioning land among villagers, and partly for draining clay soils to make them more easily manageable. It was largely concentrated in the English Midlands, though it also occurred in some northern counties and the lowlands of Scotland and Wales. Modern archaeological and historical research

Fig. 54 Medieval ridge and furrow at Upwood Meadows National Nature Reserve. Notice the reverse S shape shown by the snow lying in the furrows. This shape was needed for turning the plough with teams of oxen or horses. In summer, the better drained ridges are characterized by meadow buttercup, green-veined orchids and ant-hills, while creeping buttercup, lady's smock and rushes grow in the hollows. (Photograph T. C. E. Wells.)

has now removed much of the uncertainty that existed until quite recently about the origin of ridge-and-furrow, and anyone interested in this subject should read David Hall's scholarly little book. The accumulated evidence points to an 8th or 9th century origin, though ridge-and-furrow continued to be produced up to Victorian times when steam ploughs were used. Subsequently, the land was turned back to grass, and, in some cases, has remained in this condition for several centuries so these sites are now good examples of semi-natural habitats. Today, they often have special conservation value for their rich assemblages of plants including conspicuous species such as ox-eye daisy, cowslip, buttercups and orchids. This floristic diversity results from three factors – variability in soil conditions, consistent management over long periods, and the non-use of fertilizers and herbicides.

The strips vary considerably in size but are commonly some seven yards (6.4 metres) from crest to crest and from one to three feet deep (0.3–0.9 m) from crest to trough. On low-lying ground and heavy soils, water tends to lie in the hollows during the winter, often for weeks at a time. The ridges, on the other hand, are better drained, and the soil also tends to warm up more quickly especially on south facing slopes. The pattern of soil conditions, repeated across a field, allows one to examine the effects, as shown in the following two examples, both from Sites of Special Scientific Interest.

The first example is an elegant demonstration of the segregation of three closely related species in a ridge-and-furrow meadow made by J. L. Harper

Table 8 Plants recorded on 12 pairs of adjacent ridges and furrows at Wendlebury Meads, Oxfordshire in May 1986. Numbers and species in 0.25 sq. m. quadrats. (Original data B.N.K.D.)

Ridges		*Furrows*
Average no. species	15.2	4.7
Total no. species	40	14
No. of restricted species	31	5
Examples		
Betony *Betonica officinalis*		Cuckoo flower *Cardamine pratensis*
Dyer's greenweed *Genista tinctoria*		Meadow Fescue *Festuca pratensis*
Adder's-tongue fern *Ophioglossum vulgatum*		Soft rush *Juncus effusus*
Green-winged orchid *Orchis morio*		Creeping jenny *Lysimachia nummularia*
Greater burnet *Sanguisorba officinalis*		Creeping buttercup *Ranunculus repens*

at Pixey Mead, near Oxford. This meadow was flooded most winters and had been continuously mown for hay for at least 800 years. All three of the common grassland buttercups occurred here, meadow buttercup *Ranunculus acris*, creeping buttercup *R. repens* and bulbous buttercup *R. bulbosus*. A typical strip across the ridges and furrows showed that bulbous buttercup occupied the highest ground whereas creeping buttercup flourished in the bottoms of the furrows, and meadow buttercup was most abundant along the sides of the ridges. Where the furrow was particularly prone to flooding, creeping buttercup was 'pushed up' the sides of the ridges, meadow buttercup formed bands along the tops, and bulbous buttercup was absent. In well-drained conditions, the reverse was found – meadow buttercup replaced creeping buttercup in the furrows, and the tops of the ridges were free of buttercups altogether.

The second example is from Wendlebury Meads, also in Oxfordshire. Here, the soil is derived from a shallow layer of fine loamy drift overlying impermeable Oxford Clay. The whole area is low-lying with a high water table and subject to prolonged flooding in the furrows. The meadows have been grazed or cut for hay for 50-100 years or more. A sample survey of the vegetation across the centre of one field in May 1986 showed striking differences in floristic richness between ridges and furrows (Table 8). Not only were most of the rarer species confined to the ridges but the average number and total number of species there were significantly higher than in the furrows. How far the floristically distinct communities correlate or interact with distinct soil characteristics remains a topic for study.

The tendency towards agricultural improvement of soils has been mentioned several times in this chapter. Most ridge-and-furrow grassland has been 'improved' in the past 40 years through the use of herbicides and fertilizers, or has been ploughed out altogether. Occasionally, the ghosts of ridge-and-furrow may appear briefly in an arable field after a fall of snow, when small residual differences in soil conditions cause the snow to melt in streaks. These cultivated soils form the subject of the next two chapters.

8

Agricultural Soils: Productivity and Management

The King of Brobdingnag gave it for his opinion, that whoever could make two ears of corn or two blades of grass to grow upon a spot of ground where only one grew before, would deserve better of mankind, and do more essential service than the whole race of politicians put together.
J.D.Swift (1726) – *Gulliver's Travels*

The history of agricultural productivity in this country has been an unrivalled success story during the past 40 years. The increased scientific understanding of soil conditions and crop growth has underpinned a wide range of technological developments – new crop varieties, new machinery, better drainage, artificial fertilizers and pesticides – which have together led to crop yields unimagined in the early decades of this century.

Consider wheat yields in England, for example. We can look back over seven centuries of records, and see how average yields have gradually risen from 4.3 cwt an acre (0.54 tonnes per hectare) around AD 1200 to about 20 cwt an acre (2.5 t/ha) in 1950, a nearly five-fold increase. Since then, however, within a single farming generation, wheat yields have almost trebled again. In 1982 they stood at 50 cwt an acre while 1984 produced a record harvest at nearly 3 tons an acre (7.3 t/ha). The good conditions for root development, the uniform rain distribution and favourable day/night temperatures could scarcely have been more atuned to the growth of wheat if they had been computer- controlled. As one Cambridgeshire newspaper put it, "This almost ideal pattern certainly made a major contribution to the Latenbury Farming Company's heavy yields of generally high quality grain, averaging around...70 cwt per acre for winter wheat".

The weather in 1984 was exceptional in allowing this expression of crop potential – a near optimum uptake of nutrients from the soil with little let or hindrance from pests and diseases. Such average yields have yet to be exceeded, and it is not clear at present whether there is sufficient genetic potential for yields to rise much further. Nevertheless, even this level of production far exceeds what G.W.Cooke thought possible in 1969 when he addressed this question in a symposium on *The Optimum Population for Britain*: "The 'target' assumed in this paper is that of growing enough food for the present population. No doubt our population will be larger in 30 years time, but to grow all we need now is a difficult enough task."

Our present achievement in agricultural productivity is the result of a single-minded national policy since the 1940s to become more self-sufficient;

a reaction to the nightmare threat of starvation following the sinking of merchant shipping during the war. Agriculture is now the victim of its own success. The suggestion today is that 25 per cent of agricultural land will not be needed in the near future. In the 1989 Dimbleby lecture, HRH the Duke of Edinburgh expressed the changed position well: "It is an irony that, with the help of financial incentives, the more industrialized countries are producing surpluses of food and steps are being taken to limit output, if only for the time being."

With such a revolutionary concept dominating our vision, it is difficult to recall the concerns of the not too distant past. We no longer believe, as Sir Albert Howard did in 1940, that we may be using up the inherited wealth of our soils in the same way that we are exploiting fossil fuels. In *An Agricultural Testament*, he wrote "The restoration and maintenance of soil fertility has become a universal problem ... the slow poisoning of the life of the soil by artificial manures is one of the greatest calamities which has befallen mankind." Few people have taken so extreme a view as Howard but many have wondered whether farmers were disregarding long-term husbandry in the pursuit of short-term gains.

A run of unfavourable weather often shows up weaknesses in farming practice. Farmers in the East Midlands may still remember their dismay at seeing fields of ripening corn standing two feet deep in water after the torrential rains of July 1968. Such a disaster may be considered an act of God but the crop failures in 1969 were widely interpreted as the result of human error – of pushing land beyond its capability. As a result, the Agricultural Advisory Council was asked to undertake an urgent enquiry as to whether "the inherent fertility of the soil was being eroded and the fundamental structure of the soil damaged beyond repair". More specifically, public concern centred around the increasing use of heavy machinery, the long-term effects of continuous cereal growing, the effects of pesticides upon beneficial soil organisms (and wild life), and soil erosion.

As in so many other spheres of life, it is the increasing pace of change that creates uncertainty and concern. We now have little time to assess the effects of one innovation before it is overtaken by another. Winning food from the soil has been an uncertain business for most of agricultural history so the tried and tested methods of the past were usually the best. We can recognize some significant changes in cultivation practice over the centuries but these were the exception rather than the rule. Some innovations reflected social patterns, such as village settlements and the availability of labour. Some were in response to major climatic shifts like the period in the 8th century when the north coast of Scotland exported grain to Scandinavia. Few changes, until the 18th century, were due to technical invention or scientific insight.

A brief history

For perhaps 2000 years, agriculture maintained simple crop rotations with a fallow every alternate year or one in three on the better soils. This routine became incorporated into the medieval 3-field system which also introduced ridge-and-furrow, as described in the last chapter. During the 14th and 15th centuries, there was a general swing to pasture and sheep farming when much arable land would have gone down to permanent grass. Subsequently, some of this land went under the plough again to participate in the 4-course rotation introduced by 'Turnip' Townshend in 1730. This rotation was developed on the lighter soils of East Anglia, and alternated turnips and clover between corn crops. It was widely adopted for it allowed continuous cropping for the first time while still maintaining soil fertility. Sheep were folded onto the root crops in winter and onto the clover in summer. On the heavier Midland soils, the 4-course rotation often retained one fallow year, and used beans or peas in place of clover. Sheep were kept separate on the permanent pasture.

This farming system lasted for the next 150 years, until corn prices fell in the agricultural depression of the 1880s. Grass as a rotational crop then became widely used as a soil conditioner, and the Midlands, which had become the corn belt of England, became predominantly grass again.

Meantime, land drainage schemes had opened up new areas of fenland to agricultural use: 300,000 acres around Ely and 95,000 acres around Whittlesey – the South and Middle Bedford Levels – in the first half of the 17th century; 10,000 acres in Holland and Kesteven in the mid-18th century; 20,000 acres of Kings Sedgemoor in Somerset in the early 19th century. As the peat dried out it shrank, and natural drainage was no longer possible so hundreds of wind-driven scoop mills were built between Lincoln and Cambridge to improve the area. In one of Cobbett's rural rides in 1830, he described a fenland scene as "twenty thousand acres of land around, covered with fat sheep, or bearing six quarters of wheat or ten of oats to the acre without any manure" (6 quarters/acre is about 0.2 t/ha). The introduction of steam-driven pumps improved still further the arterial drainage in the Fens. The last great mere at Whittlesey was drained by a centrifugal pump demonstrated at the Great Exhibition in 1851. This area was thus transformed in a few years from agriculturally valueless wastes – the haunts of bittern and wildfowl – to some of the richest land in the country.

A little earlier, in 1831, James Smith published his *Remarks on Thorough Draining and Deep Ploughing*. His attention was focussed on the clay lands as he realized the importance of soil texture to water movement and successful drainage. Clay tiles were in common use by then, and infield or under-drainage made great progress, especially under the influence of Bailey Denton. In 1855 Denton made a country-wide survey of drainage need, and concluded that nearly 15 million acres in England and Wales could be improved. Modern estimates suggest that 12 million acres were in fact drained by 1880, a little under half a million a year.

Twentieth century technology, and the scientific study of soils, opened the flood-gates to experimentation and change in crop husbandry. Fertilizers were a major force as they offered a release from the centuries-old necessity for rotational cropping as the means of maintaining soil nutrients – though the early farmers would not have thought of it in such terms. The need to maximize food production during and after the second world war provided new stimulus to realize greater productivity from soils. Machinery provided ever more agricultural muscle, tractors and combines first allowing farmers to cultivate and harvest larger areas more quickly, and then requiring field sizes to be enlarged to match their maw. The ploughing up of pastures once again, the explosive growth of the pesticide industry, and the dramatic return to an almost pre-enclosure landscape in eastern England, all followed in a few decades.

This very brief historical perspective brings us back to modern times. It provides just the barest outline of the forces that have acted upon the agricultural soils of this country, and we must now consider some of the factors in more detail. We will then be better able to gauge whether the changes wrought by agricultural 'extensification' in the future will involve anything truly novel, as far as the soil is concerned, or are merely part of a longer cycle of changes which have been experienced before.

Cultivation

Ploughing is one of the oldest agricultural operations and has altered relatively little except in speed since medieval times. The ploughman's skill in turning a straight and even furrow was tested in the annual ploughing competitions that formed a regular feature of village life until recent times. How much more of a challenge this must have been when the plough was drawn by oxen or horses, and when every plough was hand-made.

Drilling seed into the soil, on the other hand, instead of broadcasting it, only dates back to the 18th century – about the time that the first factory-made ploughs were produced in Scotland. A simple drill was invented by Jethro Tull who also conceived a theoretical basis for cultivation. In his *Essay on the Principle of Tillage and Vegetation* in 1731, he put forward the idea that the sown plants competed with each other and with weeds for nutrients in the soil. Cultivation, he thought, was beneficial not only by reducing weeds but by increasing the 'pasture' upon which the roots of crops could feed. The first point was quite right but the second was a misconception. The idea that a fine tilth at the soil surface was intrinsically good for crops persisted for 200 years until E.W.Russell's experiments in the 1930s established the real purposes and advantages of cultivation. We now know that good soil aeration is important for root growth but that this is achieved by good drainage below rather than by pulverization at the surface.

The free movement of water and air through the soil depends on good pore structure and is greatly reduced by compaction. Tull noted the bad effects of compaction, remarking that the frequent treading of moist soils by horses' hoofs made the ground like a highway. He suggested fitting subsoil tines to ploughs to break up the pans so caused. The problem of compaction has always haunted farmers, and has merely taken on a new dimension as horses have given way to machinery.

In 1918, A.Amos expressed the view that heavy tractors "may do untold damage" to soils especially when they were wet. Heavy steam tractors of his day weighed 3-5 tons and could indeed make bad ruts with their metal wheels. Modern tractors are much more powerful for a relatively small increase in gross weight, but anything above 12 tons is likely to cause appreciable compaction. Laden trailers and equipment for spreading lime or fertilizer may weigh up to 15 tons and pose special risks of compacting the subsoil (below the normal depth of cultivation) if they bear on it directly through ruts. Whereas inadequate compaction in a sandy soil can be easily rectified by rolling, too much compaction in a heavy soil can take years to correct.

There are many different factors that determine soil compaction by agricultural machinery which can only briefly be mentioned here; the Scottish Institute of Agricultural Engineering has made a special study of this topic. One set of variables relates to the vehicle and one set to the soil. For a given load, one can reduce the contact pressure on the ground either by increasing the width of the tyre or by decreasing its inflation pressure. However, if the load is increased then the tyre pressure needs to be increased to support it, and this will extend the depth to which the stress is communicated in the soil. Increasing the speed can also reduce compaction because a soil can spring back through natural resilience if it has to carry a load for only a short time. Here again, though, higher speeds for a given load may require greater inflation pressures which will counteract the advantage. An institute can set guide lines for the best combinations of load, tyre width, inflation pressure and speed, but translating these into practice is another matter. The fitting of dual or cage wheels is a partial solution, and another is to use tracked vehicles. These are popular in Russia but not in this country because they are relatively slow and cannot be used on roads.

It may be necessary to traverse a field several times in a season to plough, cultivate and drill, to spread fertilizers or lime, and to spray with weedkillers, fungicides or insecticides. A good farmer knows the subtle differences that every field can have in its 'timeliness' for cultivation, but nowadays he may be constrained by the need to spray against wild oats or take-all disease at very precise times notwithstanding whether the soil conditions are good or not. The use of tramlines was introduced from Germany relatively recently; that is, the re-use of the first set of wheel marks for subsequent passes. The original purpose of this was to ensure accurate spraying, without gaps or overlaps, but it has since been adopted for cultivation as well as it greatly reduces the amount of overall compaction. This is because the first pass causes 70% of the compaction produced by several passes. The benefit to the rest of the field greatly outweighs the loss of crops in the tramlines.

Soils containing a high proportion of clay or silt are particularly prone to compaction when wet. The basic fabric of the soil is something the farmer has to live with but he can greatly reduce the risk of compaction by improving the drainage. He can also improve the resilience of the soil by introducing a grass ley into the rotation to increase the organic matter content – a return to 19th century husbandry.

The main function of ploughing, as Russell showed, is to control weeds, but ploughing also loosens the soil and creates the need for one or more cultivations before drilling. In the past, this often meant a delay of 4-6 weeks, at least on heavy land, so it was only possible to drill some of the fields on a farm before winter set in. If the winter was wet, and the land was not well drained, considerable difficulties and delays could be expected in the spring which would be reflected in lower yields. This happened in 1969 after the exceptionally wet conditions of the previous summer and autumn mentioned earlier. In terms of energy, ploughing is also very demanding and therefore expensive; it has been calculated that ploughing a nine inch deep furrow turns 1000 cubic yards of soil for every acre – 1900 cubic metres per hectare.

Really heavy, 'four-horse', land would always have been problematical for the farmer; the economic returns on his labour would be lower than on more easily worked land. This applies even in more recent times. A vivid example of the relative cost of cultivation was seen in about 1970 in the Woolley area of west Cambridgeshire where two crawler tractors were harnessed together to draw a plough: maximum traction with minimum compaction.

The introduction of heavy cultivators or chisel ploughs, instead of the conventional mouldboard plough, was a partial answer, but it was the bipyridyl weedkillers paraquat and diquat that opened the way to a new farming revolution in the late 1960s. These chemicals killed all green foliage but were inactivated on contact with the soil. They therefore made it feasible to eliminate competition from most weeds and self-sown crops after the last harvest by spraying, and then to drill directly into the undisturbed stubble. Perennial weeds, particularly couch grass, were still a problem because they could regrow from their strong underground rhizomes after the tops had been killed by paraquat. However, the next generation of weedkillers included glyphosate ('Roundup'), which was extremely effective in controlling these.

The pay-off for this reduced tillage was in allowing a major move away from spring-sown crops and grass to winter cropping. Much larger areas could be sown with higher yielding, and much more profitable, winter cereals. The extra money in the farmer's pocket in the 1970s was invested in better drainage and better machinery – a spiral of improvement. There were also some incidental benefits in the early years. On heavy land, natural drainage and root development improved after direct drilling because of the greater continuity of pores and channels that became established through worm activity (*cf.* chapter 5). On light land, the risk of soil erosion and the loss of organic matter was reduced.

By the early 1980s, tine cultivators, chisel ploughs and direct drilling had replaced the traditional mouldboard plough on about 50% of the heavy land in the drive towards quicker, more energy saving and more profitable agriculture. In farming terms, this represented a major revolution compared with 10–15 years previously when the mouldboard plough reigned supreme. However, the new farming practice had scarcely become established in the popular consciousness before its heyday was already past.

There were serious disadvantages in reduced tillage that became evident after a few years. In many cases, but particularly on light soils, the lack of ploughing led to soil compaction below the shallow working depth. This was especially noticeable on headlands and near field entrances, and led to substantial yield loss. The Agricultural Development and Advisory Service (ADAS) now recommends loosening the soil every few years to avoid this problem. Perhaps the major drawback of reduced tillage was the increase in grass weeds in the growing crop, especially blackgrass *Alopecurus myosuroides* and bromes; these are more difficult to control with selective weedkillers than wild oats. The cumulative increase of bromes through the carry over of seed from one year to the next was a major cause of the decline in direct drilling.

Fertilizers

Some 15 to 20 chemical elements are essential for plant growth, of which crops need 9 or 10 in quite large amounts and the remainder only in trace amounts (Table 9). Of the major elements, carbon, hydrogen and oxygen are obtained from the air and water (as CO_2 and H_2O) while the others are mainly taken up by the roots from the soil in the form of 'available' nutrient ions (chapters 1 and 3). Thanks to atmospheric pollution, more than enough sulphur is supplied in 'acid rain' and dry deposition to meet the needs of most crops except in parts of Scotland. However, these aerial inputs have been declining – from 57.7 to 36.4 kg/ha between 1979 and 1982 at Hurley in Berkshire – and are likely to be reduced still further to overcome the undesirable effects of suphur dioxide on the environment. In the last few years, sulphur deficiency has been found in grass for silage in Wales and south west

Table 9 Minor nutrients needed by crop plants (from Davies, Eagle & Finney, 1982).

	Deficiency situations		
Element	*Crops*	*Soil types*	*Conditions*
Manganese	cereals potatoes sugar-beet	sandy soils lime-rich peats	excess lime
Iron	sugar-beet pears	chalk	excess lime
Copper	cereals	light peats heathland sands humus-rich chalk	acid conditions high organic matter excess lime
Boron	brassicas beet, swede	sandy soils	excess lime drought
Molybdenum	cauliflower	rare in Britain	acid conditions
Zinc		not a problem in Britain	excess lime

England, and one case was reported in barley in Wales during 1988. We may, therefore, have to supplement the supply in the future.

Sodium, the cation of common salt, is only needed in large amounts by crops of maritime origin such as mangolds, spinach and sugar-beet. Calcium is important for keeping the soil 'sweet', i.e. in preventing soil acidity. Its use goes back to Roman times but it is rarely in short supply as a nutrient. Most farm land in Great Britain still supplies the 5–25 kg/ha of magnesium required by crops, though deficiencies are increasingly common on lighter soils.

The three elements, then, that are most likely to limit crop production are nitrogen N, phosphorus P and potassium K, and of these nitrogen is the key to high yields. Most of the nitrogen in soils is fixed and cannot be used by plants. The nitrogen in organic matter is also unavailable until it is converted to the inorganic nitrate (NO_3^-) and ammonium (NH_4^+) ions (chapter 6). A hundred tonnes of organic matter per hectare, accumulated under a grass ley, contains substantial quantities of N, about 200 kg of which will be gradually released in the 2–3 years after ploughing. Primitive agriculturalists practised shifting cultivation as a way of cashing the store of N and other nutrients accumulated under natural plant communities without re-investing anything; they simply moved on when the soil became too poor to sustain further crops. Settled agriculture, on the other hand, has to balance the nutrients extracted and lost from the land with the rate of replenishment. Natural replenishment comes from the weathering of rock particles and the input from rain. Nitrogen is also captured from the air by soil algae and some bacteria, especially those associated with the root nodules of legumes as described in chapter 3. The use of clover was, therefore, an important step forward. Farmyard manure (FYM) was carefully conserved and recycled, and gradually other organic and inorganic substances were introduced, such as ground bones, guano and saltpetre.

By the early 19th century, the chemical composition of these 'natural' and 'artificial' fertilizers was known: saltpetre, for instance, is potassium nitrate (KNO_3); guano from fish-eating birds contains mainly phosphorus and ammonium. The problem, in the early days of scientific agriculture, was to derive unambiguous results from their use. During the 1840s, various experiments were done with combinations of manures and mineral salts on single plots, but the results were inconsistent and even contradictory. J.S.Henslow at Cambridge, J.F.W.Johnston at Durham, and C.G.B.Daubeney at Oxford all saw the need for organized experiments with single substances. Such experiments, they felt, should also include the use of two untreated plots in order to judge the variation due to natural causes – the beginnings of statistical design! If barley produced a heavier yield from one plot receiving substance X than from another plot without it, could the difference be attributed to substance X?

In hindsight, it is easy to see the need for replication and for 'control' plots to take account of other factors such as local variations in soil fertility, drainage, pests or disease. At the time, such influences could only be guessed at, and it was J.B.Lawes in 1847 who most clearly defined the problem for agriculture by asking "What substances is it necessary to supply to the soil in order to maintain a remunerative fertility?" And it was he who, with J.H.Gilbert, decided "to make experiments at once more systematic and on a larger scale on some of the most important crops of our rotations." Thus was laid down the most famous of all long term agricultural experiments with the growing of wheat on Broadbalk field at Rothamsted (Plate 12). This experiment, and another for root crops at Barnfield, continue to the present day.

In addition to unmanured and FYM plots, there were two main groups of treatments. One group tested the effects of phosphorus, potassium, sodium and magnesium separately and in various combinations in the presence of nitrogen, while the other compared different amounts and forms of nitrogen (NO_3 *versus* NH_4) in the presence of all these other four nutrients together. The yields of wheat from the Broadbalk plots have been summarized from time to time. They make fascinating reading for anyone with a statistical bent.

Some of the results are shown diagrammatically here in increasing order of yield (Fig. 55). Looking up the columns, one sees the improvements brought about simply through the introduction of better varieties of wheat during this 120 year period, coupled with better pest and weed control. It is interesting to note that yields have been maintained and even increased somewhat on the plots which have received nothing since 1843. This is the level sustainable with natural inputs alone. It is similar to the world average.

Scanning across the columns, one sees the enormous response to artificial fertilizers especially with modern wheat varieties. Recent yields from the NPK plots are a little above the national average. The last column on the right gives the yield from the use of FYM. The next chapter discusses organic farming. For the moment, it is enough to note that 35 tons of FYM contain about twice as much N, about three times as much K, and a little more P than the levels in the artificial fertilizers supplied to plots 13 and 7.

One of the reasons why the early experiments were so difficult to understand was because one cannot always predict the combined effect of two nutrients from a knowledge of their separate effects. Figure 56 shows an example in which K increased the yield of sugar-beet by about 0.14 tonnes/ha, N increased it by 0.36 t/ha, and both together increased it by 0.57 t/ha. This suggests that the individual effects are roughly additive; the dashed lines in the figure show the predicted additive effect. However, if one doubles the dose of both N and K, the picture changes: the two together 'interact' to produce much more than one would expect from their separate

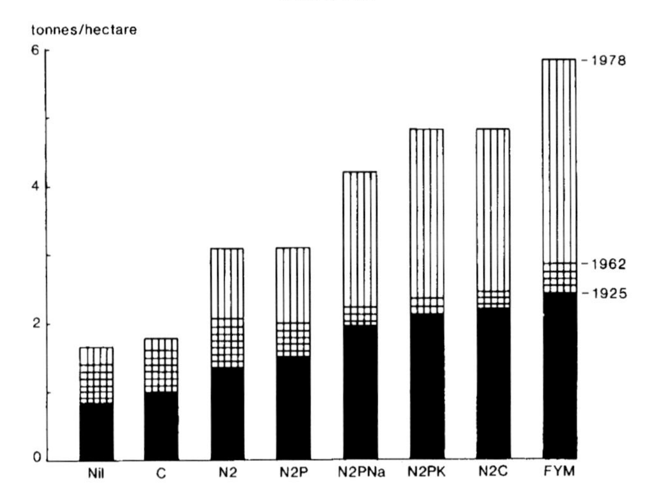

Fig. 55 The effects of different nutrient combinations on yields of wheat grain from Broadbalk, Rothamsted. Average yields between 1852–1925 (black columns), 1935–62 (cross-hatched), 1970–78 (hatched). N2 = nitrogen rate, P = phosphorus, K = potassium, Na = sodium, C = P + K + Na + magnesium, FYM = farmyard manure. (Data from *Rothamsted Annual Reports* for 1968 and 1982, plots 3, 5, 10, 11, 12, 13, 7, 22.)

effects. The third figure shows an even greater interaction when N is combined with a cocktail of other essential nutrients in increasing the yields of potatoes.

In practice, it is possible to build up the reserves of P and K in the soil with fertilizers because they are adsorbed by soil particles and then gradually released to crops with little loss. Inorganic N, on the other hand, is like quicksilver. If it is not captured by plant roots soon after it is applied or mineralized from organic matter, it is liable to slip away down the soil profile and into the ground water or out through the field drains (see next chapter).

Although the basic concepts of crop nutrition have been known for a hundred years, farmers themselves regarded artificial fertilizers with suspicion until the second world war. The County War Agricultural Executive Committees met with considerable opposition in their attempts to persuade farmers to use more fertilizers. It was only experience – and observation of their neighbours – that convinced them that fertilizers could improve yields dramatically without ruining soils[1].

---

1 Traditionally in Britain, fertilizers have been measured in terms of nitrogen N, phosphate P_2O_5 (which is actually phosphorus pentoxide), and potash K_2O. One pound of phosphorus was therefore expressed as 2.29 lb of phosphate, and 1 lb of potassium as 1.20 lb of potash. Fertilizers used to come in hundredweight bags, and 1/100 cwt or 1.12 lb was defined as a 'unit'. Thus: 1 unit of nitrogen = 1.12 lb N, 1 unit of phosphorous = 1.12 lb P_2O_5 = 0.49 lb P and 1 unit of potash = 1.12 lb K_2O = 0.93 lb K. Many combinations of compound fertilizers are available e.g. high N with low P (25:0:15) or low N with high P and K (9:25:25). A hundredweight bag of 20:10:10 fertilizer contained 20 units (22.4 lb) of nitrogen, 10 units of phosphate (4.9 lb P) and 10 units of potash (9.3 lb K). In metric terms, 1 cwt is roughly equal to 50 kg, 2 units to 1 kg and 0.8 units/acre to 1 kg/ha.

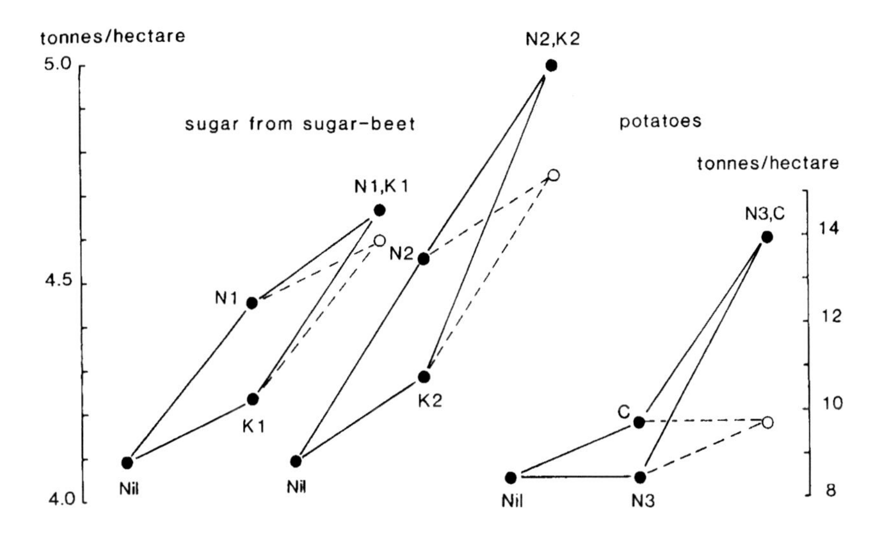

Fig. 56 Examples of 'interaction' effects of fertilizers on yields. Each figure shows the increase in yield with one or other or both nutrients, the dotted line indicating the expected yield with both nutrients if their effects were simply additive. The first pair of figures shows little interaction with low levels of nitrogen and potassium but large interaction with doubled levels. The third figure shows no effect on yield of potatoes by adding nitrogen on its own but a large effect when combined with P + K + Na + Mg. (Sugar-beet data from G. W. Cooke 1982; potato data from Rothamsted Annual Report for 1982.)

After a rapid increase in use, the amounts of P and K applied to soils have stabilized since 1956. However, the use of N in the UK climbed steeply and continuously from about 61 thousand tonnes in 1939 to 1450 thousand tonnes in 1983. Rates of use continued to increase until 1986 but then levelled off and have even shown a small decline in 1988. These overall figures, of course, disguise big differences between crops, and trends within them, as acreages (or hectareages) grown, market prices and subsidies have changed. From a very small base in 1970, oil-seed rape, for instance, has increased enormously in extent and now receives more nitrogen per hectare than any other crop – 254 kg/ha in 1980 on 93,000 ha. As shown throughout this chapter, present-day yields are the sum of many independent developments. Potential yields, though, cannot be realized without large inputs of fertilizers or manures, as Figure 55 makes clear. It is possible, therefore, to calculate how much production would be foregone without them. In the case of cereal grain, this amounts to 9.3–10.5 million tonnes per year during 1982–3. At annual costs of fertilizers of £180–300 million, this gives net benefits to UK farming of £536–685 million. Some of the hidden costs are discussed in the next chapter.

Grassland, too, has become a major recipient of N. No longer is the contribution from clover enough to meet the demands. Most lowland pastures, indeed, contain very little clover today. Nearly a third of recently re-sown pastures now receives at least 250 kg/ha of N to maintain production, with corresponding high inputs of P and K. Scientists at the Grassland Research

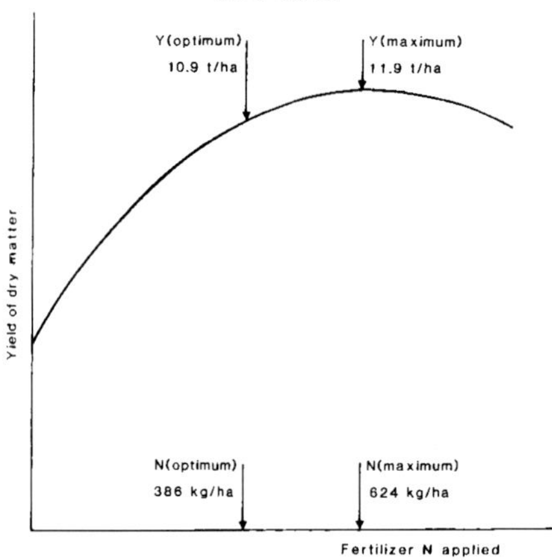

Fig. 57 General form of the response of grass to fertilizer N showing the key points affecting upper limits of the response. The optimum yield, Y (optimum), is defined as the point where the response to N falls to less than 10 kg of dry matter per kg of N applied. The mean values given are based on 21 sites in 20 counties. (Redrawn from J. Morrison and others 1980.)

Institute have determined the optimum, value-for-money, yield of grass at about 11 t/ha, for an input of 386 kg/ha of N. Another tonne could be grown but only by greatly increasing the amount of N (Fig. 57). The real secret lies in the timing and the doses. By giving applications at 3–4 week intervals between March and September, the maximum amount of N (50–60%) can be taken up from the soil; and by giving the highest dose in midseason, a more even pattern of production can be achieved without reducing total output.

In the 1984 Macaulay Lecture, Sir Leslie Fowden, former director of Rothamsted, described the current approach to the use of nitrogen fertilizers. This included the development of a computer-based model which would simulate the movement of N in the soil. Given a knowledge of soil temperatures and rainfall, the computer could work out the mineralization of organic N, its movement down the soil profile, and its uptake by crops. "Soon", he said, "farmers may be able to adopt such a model, and, with their own computers, introduce parameters descriptive of their own local situation and target yields to predict the optimum spring-N applications for individual fields". On present form, we are, in fact, unlikely to be able to predict the total inorganic N pool in the soil within reasonable limits from year to year so this vision still eludes us. In the radically altered agricultural scene today, such an objective may, indeed, seem less pressing.

Field drainage and irrigation

Badly drained fields create several kinds of problems for farming. The most obvious one is the danger of damaging the soil structure, by trampling in grassland (known as poaching), or by the passage of vehicles in arable fields. If the ground is too soft, it can delay drilling in the spring, and seriously limit the ability to spray against weeds and diseases at critical times. A wet soil also needs more nitrogen fertilizer than a well drained one to produce the same effect on crop growth. And finally, the roots of most arable crops and fruit

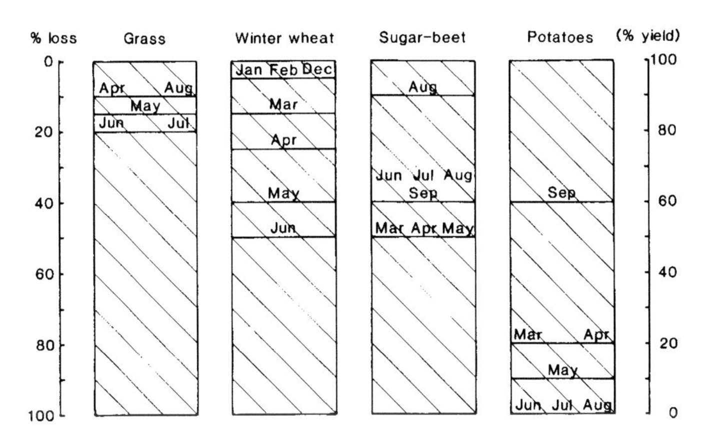

Fig. 58 Percentage loss in crop yields due to 7 days waterlogging occurring in specific months as shown. Waterlogging in other months has no effect. (Data from B.D.Trafford 1970.)

can suffer badly if there is an excess of water for any length of time, especially during the growing season.

This last effect was nicely shown by some Hungarian work on various crops subjected to waterlogging for 3, 7 and 15 days in each month of the year. Grass, as one would expect, was least affected; that is why grassland predominates over arable land in areas of higher rainfall. But even with grass, the yield was halved if it experienced a fortnight of waterlogging in May, June or July. Potato tubers drown easily, and even one week's waterlogging between June and August could lead to the total loss of the crop. Winter wheat was intermediate but could suffer substantial losses from prolonged waterlogging during the winter months; which would not be unlikely on poorly drained soils. These results are perhaps not exactly transferable to British conditions but they are a good guide (Fig. 58).

Partly because of the loss in crop vigour that occurs in badly drained soils, the competition from weed grasses is increased, and diseases can take hold where they could not on healthy plants. Poorer crops also extract less water from the soil in transpiration, and so it does not dry out so quickly. These various effects exacerbate each other, thereby reducing productivity and thus the income which the farmer has to invest in improving soil conditions.

More than half the agricultural land in Britain needs artificial drainage to shed excess rainfall and maximize crop yields. A survey in 1969 suggested that a quarter of the 27.3 million acres (11 million hectares) of agricultural land in England and Wales had already been drained, and another quarter would benefit from tile drainage. The Ministry of Agriculture, Fisheries and Food (MAFF) offered grants of 50–60 per cent towards the capital costs of approved drainage schemes during and after the war, with the result that there was a sustained effort for over 40 years. Rates of drainage installation increased from about 15,000 acres a year in 1944 to 105,000 acres a year in 1960, and then levelled out at about 250,000 acres a year between 1970 and 1980. At this rate, it was estimated that all the land thought to need draining in 1970 would be drained by 1990. However, drainage grants have now

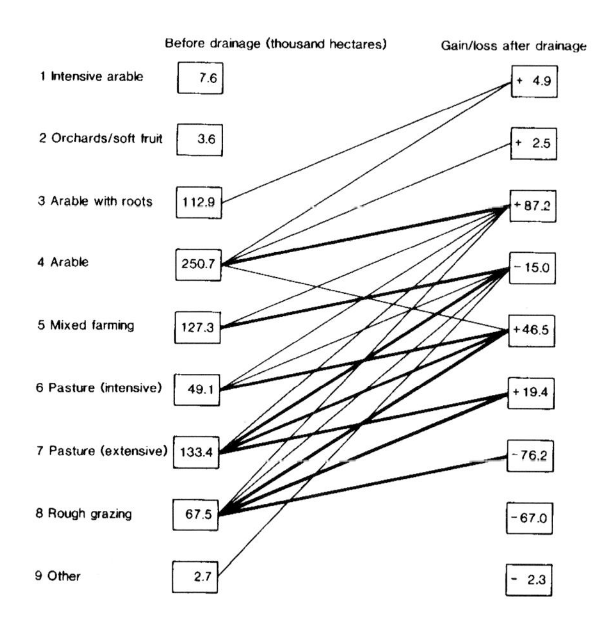

Fig. 59 Main changes in land use as a result of drainage during 1971–1980 in England and Wales. Light lines indicate changes of 1000–10,000 hectares, heavy lines indicate changes of 10,000–71,000 hectares. (Data from MAFF Drainage Statistics 1978–80.)

virtually ceased, with the change in agricultural emphasis towards reduced rather than increased output, so this target may never be reached.

The optimum depth and spacing of tile drains depends on the permeability of the soil and the slope of the land as well as on the requirements of different crops. High value crops like potatoes need drains 10–20 metres apart on level ground. On the other hand, cereals or oil-seed rape may only need drains 200 metres apart on sloping ground because water can be fed into them by mole drains 2–3 metres apart, like weft over warp. Moling is cheap, and the tunnels can last several years before collapsing or becoming choked.

It is difficult to measure the benefit of drainage precisely – to estimate what B.G Davies calls 'drainage worthwhileness'. Any experiment on classical lines needs large areas with and without drains, but comparable conditions over large areas are difficult to achieve so one cannot be sure whether other variables influence the results.

One such experiment in Cambridgeshire measured yields of winter wheat for six years. The average yields from plots with drains was 33.7 cwt an acre (4,230 kg/ha) while for the undrained plots it was 8.5 cwt an acre less; in 1960 this would have been worth £8 10s an acre. Another experiment, by the Field Drainage Experimental Unit of MAFF which was set up in 1962, measured liveweight gain in bullocks grazing on drained and undrained grassland in Devon. There was a noticeable difference in weights, and therefore in financial returns. (The market value of beef at the time was 1s 6d per lb!). However, the capital cost of drains, at 100 ft spacing plus subsoil cultivation, was

£75 per acre, so it was doubtful if drainage would have been a worth while investment for such a low value enterprise.

What was more likely was that the farmer would change to a more intensive form of land use which drainage made possible, and which gave higher yields. The Ministry of Agriculture have classified land use under nine categories depending on the intensity of use, and have recorded marked shifts in land use as a result of drainage between 1971–1980. These are summarized in Figure 59 which shows the especially large gains in categories 3 and 5 – arable with roots, and mixed farming – largely at the expense of extensive pasture and rough grazing. For example, of the 4268 hectares of rough grazing in the south eastern region which were drained during this period, only 28 hectares remained as rough grazing; and similarly in Wales, only 149 hectares remained in this category after draining 21,041 hectares.

The draining of moorland in Britain has likewise been encouraged through 50 per cent grants since the 1940s, and by 70 per cent grants under the European Community's Common Agricultural Policy for improving the productivity of hill sheep in 'less favoured areas'. No drainage statistics have been collected for hill land, however, and there seems to be little hard evidence that the land has become markedly more productive for sheep, red grouse or other game.

Open ditches are dug at 15–35 metre spacings but water is held so well by the peat that the effect on the water table and on vegetation is confined to a few metres on either side of the ditches. Afforestation may be the best way of lowering the water table through transpiration, but this changes the land use entirely, and has other effects on the soil.

Irrigation is the other side of the coin from drainage. On a hot sunny day in Britain, a crop will draw about 10 tons or 2240 gallons of water an acre from the soil in transpiration (about 25,000 litres/ha). If this water is not replaced, the crop may eventually wilt and finally die. The optimum condition for crop growth is when the soil is at field capacity, as described in chapter 1. For each kind of crop, there is a critical soil moisture deficit below field capacity when growth and yield are reduced. This point can be reached quite quickly on light sandy soils in East Anglia which do not hold much water at field capacity, and where there is little rainfall to keep the level topped up.

Irrigation is an obvious answer, and could be more widely used to extend the range and profitability of crops. It depends, of course, on an adequate source of water and a means of monitoring the soil water deficit. An open lysimeter is one method; rather crude and expensive to install but easy to use. The other method involves keeping a water balance sheet. Here, a running tally of soil moisture deficit (SMD) is kept by entering daily rainfall, and subtracting the water lost in evaporation and transpiration (evapotranspiration), thus:

SMD (today) = SMD (yesterday) + evapotranspiration – rainfall

Agrometeorologists can now provide weekly 'potential evapotranspiration' figures for any particular district based on weather data from the nearest weather station. The 'actual evapotranspiration' figures used in the equation above are then calculated from a knowledge of the particular soil – its moisture holding capacity and so on – and the state of growth of the crop concerned. For irrigation purposes, a plan might specify, for example, that 25mm of water should be applied whenever the calculated soil moisture

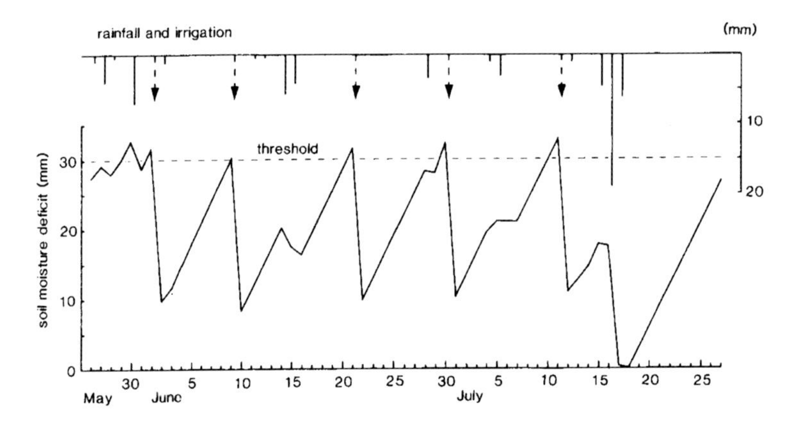

Fig. 60 Example of changes in soil moisture deficit and irrigation needed for a potato crop over a period of 9 weeks. The arrows indicate the times when the soil moisture deficit exceeds the laid down threshold value of 30 mm, and 25 mm of irrigation is therefore applied. (Data from MAFF, ADAS Cambridge.)

deficit reached 30mm. Figure 60 shows, in graphical terms, the water balance sheet for maincrop potatoes over a nine week period from the last week of May to the end of July.

During this period, the estimated daily water loss ranged from 2.7mm in May to 3.1mm in June, while the average daily rainfall for the first seven weeks was less than 0.7mm. The crop therefore had to be irrigated five times as the soil moisture deficit kept climbing remorselessly back to the threshold value. During 16–18 July, however, in the eighth week, there was 29.6mm of rain and so it was a fortnight before any more irrigation was needed. Notice that the rainfall was enough to cancel out the accumulated soil moisture deficit almost exactly, and return the soil to field capacity. Any more rain would have been assumed to be surplus and to drain away or run off, i.e. soil moisture deficit cannot be negative.

Irrigation is most valuable for potatoes, especially early potatoes, secondly for most vegetables, thirdly for sugar-beet, and lastly for cereals. Large-scale sprinklers, rain guns and giant rotary irrigators are all used nowadays (Fig. 61).

Straw disposal

Often again it profits to burn the barren fields, firing their light stubble with crackling flame. It is whether the earth conceives a mysterious strength and sustenance thereby, or whether the fire burns out her bad humours

Virgil (37 BC), *The Georgics Book 1*

Straw burning was introduced in the 1960s as a result of research by ADAS and caught on, dare one say it, like wildfire. The benefits and drawbacks have been studied ever since, and, as the implications for soil management are so great, they are worth considering in some detail.

As yields of corn have increased dramatically in the past 20 years, so have

Fig. 61 A rain gun irrigating a potato field during the dry summer of 1989. (Photograph B.N.K.D.)

the quantities of straw produced, largely in parts of the country where the number of livestock, and therefore the demand for straw, has decreased. In the early 1980s, the total amount of straw produced in the United Kingdom was over 12 million tonnes, of which more than 5 million were surplus to agricultural requirements. On individual farms, the amount of baled straw probably averaged about 4 tonnes a hectare but measurements have shown that an equal quantity is left behind. A heavy crop, therefore, could mean that there were 10 tonnes a hectare to dispose of one way or another.

There are three main options for straw disposal open to farmers. First, the straw can be baled and removed, as is usual with barley anyway, and common in the case of wheat where it can be used for bedding for stock. Secondly, the straw can be chopped up, spread and ploughed in. And thirdly, it can be burnt in swathes in the traditional way or after spreading or collecting into heaps.

Where there is no demand for straw, burning has some obvious advantages for soil cultivation: it is easy to do and cheap, it allows the soil surface to dry out quickly after showers, and helps to produce a friable texture for direct drilling. Temperatures at the soil surface reach a peak within about half a minute. They may exceed 100°C for over a minute, but the soil cools down again quite quickly as the fire moves across the field. Most of the heat is dissipated upwards, so, even with the largest amounts of spread straw, the temperature a few millimetres below the soil surface remains below 60°C.

A good burn can also halve the number of black-grass seeds that germinate. Apart from competing directly with the crop, grass weeds harbour aphids that carry barley yellow dwarf virus which have to be controlled with pyrethroid or other insecticides. Unburnt straw, whether left on the surface or ploughed in, encourages slugs which again require pesticides (molluscicides)

Fig. 62 A cereal field after straw burning, showing the harrowed perimeter strip. (Photograph B.N.K.D.)

to control them. (Fungal diseases, such as eye-spot and take-all, are not significantly reduced by straw burning, as had been thought, probably because it only needs a few live spores to start an infection if conditions favour their development.)

Ploughing the straw into the soil, on the other hand, has few advantages and several disadvantages. In the first place, it requires more energy than burning and minimal cultivation, and is therefore more expensive; it could increase net costs on heavy land by £20–50 a hectare. Straw incorporation probably improves structure and workability but this is difficult to quantify. The roots, unburnt stubble and chaff contribute more organic matter than the cut straw, so ploughing it in produces only a small increase; incorporating 5 tonnes/ha of straw into a silty clay loam for five years increased the soil organic matter by 12 percent in one trial, while other, 17-year, trials showed no significant gain. Straw contains small amounts of phosphate and potash but very little nitrogen. The high carbon:nitrogen ratio of 80:1 means that the bacteria which attack the straw need extra nitrogen which has to be taken from the soil reserves or from fertilizers applied – about 8 kg per tonne of straw or about 60 kg per hectare. It was thought at one time that this nitrogen demand, and the phytotoxic effects of breakdown products, posed serious problems for seedling development, but it is generally agreed now that these fears were overplayed.

Despite the agricultural advantages of straw burning, they have been outweighed by the environmental disadvantages. In the 1960s, much damage was done to hedgerows and trees by uncontrolled straw burning, apart from the occasional damage to neighbouring property. Bye-laws now require farmers to remove the straw from a wide strip around a field, or to plough it in, before the field is burnt (Fig. 62). Even this, however, has become less acceptable to the public at large because of the pollution from smoke and smuts. If weed problems and soil compaction from direct drilling had not already caused a return to ploughing, the objections to straw burning would probably have produced the same result. Straw incorporation is now becoming common practice, and the implications are being thoroughly studied so that farmers can adopt the best agronomic measures.

Postscript

The theme of this chapter has been soil productivity for agriculture. It has not considered the much broader issues of landscape, farm subsidies, nature conservation or the balancing of national needs. These issues have been variously addressed by other writers, such as Nan Fairbrother in *New Lives New Landscapes*, Marion Shoard in *The Theft of the Countryside*, and Kenneth Mellanby in *Farming and Wildlife*.

Even within the narrower theme of productivity per unit of land, one must remember that good soil management and fertility are only part of agronomy; the important roles of plant breeding and crop protection have not been dealt with. We now know that high input agriculture can be much more productive than was thought possible 20 years ago when G.W. Cooke discussed the question of food production in Britain. Furthermore, fears that the soil structure was being fatally damaged, and its inherent fertility was being destroyed, have not been fulfilled. In agricultural terms, the soil structure has actually been improved in many cases, as increased crop yields have been accompanied by greater crop residues ploughed back.

Of course, these improvements have been bought through increased dependence on agrochemicals and through the simplification of ecological systems. In concentrating on production for human consumption, less energy is dispersed through alternative pathways, and this must be reflected in reduced diversity of plants and animals. The next chapter considers some of the problems and indirect costs of intensive farming, including the effects of farming practices on the soil fauna, and the capability of the microflora for dealing with pesticide residues in the soil. It also reviews the risks of soil erosion and the pollution of ground water by nitrates.

9

Agricultural Soils: A Sideways Look

Chapters 4 and 5 mentioned a range of soil pests including eelworms, slugs and many insects, but what about the rest of the soil fauna? Does it make any beneficial contribution to agriculture? The possible significance of earthworms and other invertebrates was reviewed in the late 1950s and early 1960s but with rather uncertain conclusions. The activities of worms – in draining and aerating the soil, in macerating dead vegetable matter and rendering it more available to attack by microorganisms, and in helping to create good crumb structure – were evidently what Sellar and Yeatman would have called a 'Good Thing' in *1066 and All That*. It was another matter, however, to demonstrate this quantitatively. Small scale experiments, in which earthworms were added to pot-grown cereal plants, were inconclusive because, although there was some indication that worms improved yields, it often seemed that they were equally effective alive or dead!

Though this approach has proved rather intractable, there has been a great deal of productive research in Britain, especially at Rothamsted Experimental Station by C.A.Edwards and others, and also in Europe and the USA, on the incidental effects of agricultural practices upon the soil fauna. This chapter considers some of these effects and the way the microflora deals with the array of pesticides that are applied to farmland soils. It then looks briefly at other problem areas of modern intensive agriculture, and finally at organic farming as an alternative.

Cultivation

Realistic estimates of the numbers of soil organisms were just becoming available around 1950 when E.W.Russell was revising the 8th edition of Sir E.J.Russell's *Soil Conditions and Plant Growth*. Much effort had been spent in developing extraction techniques for pests such as wireworms and eelworms, and was being extended to groups such as micro-arthropods. It thus became clear just how impoverished the fauna of arable soils was in comparison with permanent grassland. Ploughing causes a drastic reduction in most groups of animals, an effect which is intensified if the soil is kept in a fallow condition. Only those species persist that are capable of living in mineral soil without a surface humus layer. They must also withstand the rigours of change: the physical upheaval of ploughing, and the greater fluctuations in micro-climate that occur under alternating bare ground and cropped soil. Direct comparisons have shown that permanent grassland supports about twice as many species as arable land.

Earthworms show the effects of the change from grassland to arable cultivation very clearly. Populations of worms are generally three times as high in grassland, and have four times the weight due to the larger proportion of big species, for example, the lob worm *Lumbricus terrestris*. When a field is ploughed, there is an initial loss of worms caused by direct physical damage and also due to predation by birds. Gulls are often seen following the plough in autumn, swooping down to pick up an exposed worm before it can bury itself again, like gulls behind a trawler picking up scraps of fish. There may also be a more gradual toll of worms because of increased drying out of surface layers of the soil and penetration by winter frosts. Probably, however, the most important factor is simply the gradual loss of organic matter as a result of continuous arable cropping, which is a relatively recent departure from traditional farming. About 50 per cent of the biomass of cereals remains behind in their roots, but this is small compared to the organic matter built up in the soil under grassland. Crops, like sugar-beet or potatoes, leave even less organic matter behind after harvest, and so reduce the earthworm population still further.

Direct drilling of cereals gives an interesting new twist to the story, since by eliminating cultivation there is a partial reversal of the trend. Eight years of continuous cereal cropping with and without cultivation allows one to compare these alternative systems of cropping. Differences in the size and composition of earthworm populations are fairly consistent and often quite striking. Generally speaking, the total worm populations decrease under a system of continuous long-term cereal cropping irrespective of whether the soil is cultivated before sowing or not. This suggests that there is a gradual depletion of the worms' food resources. However, fields that are direct drilled support larger populations on average than those that are ploughed every year. A less drastic form of cultivation such as chisel ploughing, which breaks up the soil surface but does not invert it, has an intermediate effect on the worms.

Disruption is greatest for those species of worms that produce deep, semi-permanent burrows. Thus, after eight years of regular ploughing, the long worm *Aporrectodea longa* and the lob worm *Lumbricus terrestris* are almost eliminated from some soils. Smaller, surface-living forms, such as the green worm, grey worm and rosy worm *Allolobophora chlorotica*, *Aporrectodea caliginosa* and *Aporrectodea rosea* are less affected; they decline in numbers progressively and almost equally under all forms of continuous cereal growing.

The burning of straw destroys large numbers of surface-living creatures but the effects on populations are not obvious since in any case they often experience large seasonal fluctuations. The animals are usually very mobile and able to recolonize a field very quickly. Also, many species have young stages which remain largely unscathed within the soil. Nevertheless, numbers of beetles, mites and springtails seem to decrease with repeated episodes of straw burning, especially on the large scale that has been practised in parts of East Anglia; compared with those on unburnt areas, their numbers may remain low for several months after burning. Earthworms are affected indirectly through the loss of a food resource and the rapid drying of the soil surface. A single event of this kind shows little effect but repeated burning over several years reduces worm populations.

Pesticides

During the last forty years or so, pesticides have been developed and used on an ever increasing scale. They are powerful tools in the service of man for combating a wide range of medical and agricultural pests. Ever since early agriculturalists were confronted with tares and locusts as serious competitors for the resources they won from the soil, means have been sought to limit their effects. Only with the growth of the modern chemical armoury of plant protection chemicals have we fully realized the extent of losses sustained, and the potential benefits of clean, healthy crops. As one major problem after another has been overcome, so new ones have been uncovered. The challenge from weeds, fungal diseases, eelworms, insects, mites, slugs and rodents has been met by a battery of herbicides, fungicides, nematicides, insecticides, acaricides, molluscicides and rodenticides. These chemicals permit the channelling of energy and nutrients into chosen crops instead of allowing them to be shared, as in Nature, among other forms of life in complex food webs.

Like an unpredictable genie, pesticides have proved to be a somewhat mixed blessing, for their overall effects can seldom be fully predicted. There are few if any pesticides that are completely specific to the target organisms; discrimination between harmful and harmless organisms is rarely adequate. This was particularly the case in the early days, when compounds like DNOC (dinitro-ortho-cresol) were equally effective against certain insects and some weeds, and when DDT was considered outstanding precisely because of the wide range of insect pests that it controlled. Later, this broad spectrum of activity was seen to have its drawbacks, especially after the many ramifications of its toxicity and persistence were revealed through patient research.

For instance, there was the famous case in the late 1950s of the death of robins on the campus of Michigan State University (The American robin is more like our thrush or blackbird). It appeared that the birds had absorbed large amounts of DDT by feeding on earthworms. These, in turn, had accumulated the insecticide from elm leaves months after the elm trees had been sprayed with DDT to kill the beetles that carried Dutch elm disease. A Punch cartoon of the day described this food chain in a clever paraphrase of the well-known nursery rhyme: 'This is the bird that ate the worm that fed on the leaf that Jack sprayed!' The example illustrates the point that a worm cannot afford to take too narrow a view of pesticides, for the soil is often the ultimate repository of sprays even when they are applied to crops or trees.

The controversy over pesticides in America was brought to a focus in Rachel Carson's book *Silent Spring*. At the same time, the tension produced within the international scientific community initiated a very productive period of analytical, ecological and toxicological research during the 1960s. Wild life aspects of this story have been well captured in John Sheail's book *Pesticides and Nature Conservation: the British Experience*, while a general account of pesticides, their value as well as their undesirable effects, has been admirably given in the New Naturalist series by Kenneth Mellanby.

DDT was discovered in Switzerland by P. Müller of the Geigy Chemical Company just before the start of World War II. It was in large scale use in the 1940s for the control of medical pests and saved millions of lives. It was subsequently used so widely on crops and fruit that for some years it was the most common pesticide residue found in agricultural soils. DDT and HCH (formerly known as BHC or Lindane – the active gamma isomer of hexachlorocyclohexane) were the first of a line of insecticides known as chlorinated hydrocarbons or organochlorine compounds; they were followed by compounds such as aldrin, dieldrin, chlordane, endrin, heptachlor and endosulphan. The chlorine atoms in these molecules, as well as their insolubility, renders them peculiarly resistant to microbial breakdown, so they remain effective in soil for many weeks or months. Adequate persistence is an important requirement for ensuring protection against pests which cannot be sprayed directly but which might attack a seed or plant at a later date. However, lack of selectivity results in the killing of beneficial and harmless species, and the danger of uncovering new pests hitherto kept under control by predators. Moreover, lingering sublethal persistence is a disadvantage as it may produce resistant pests.

The persistence of pesticides in soil varies widely depending on conditions, but is most marked when the chemical is mixed into the top few centimetres of soil. Around 80 per cent of DDT thus applied can be detected in the soil a year later, with 50 per cent after three years and 5–35 per cent after ten years. Compounds such as HCH and aldrin disappear more quickly but, whereas HCH breaks down into non-insecticidal substances, aldrin is oxidized to dieldrin which is equally insecticidal and more persistent. DDT has the added complication of being broken down, or partly metabolized, within the bodies of beetles, worms and slugs into DDE and sometimes TDE (rhothane) which produce sublethal effects in them or their predators.

Before 1964, aldrin and dieldrin were used in dusts, drenches and sprays and sometimes mixed with fertilizers against a wide range of soil pests including leatherjackets, wireworms, cutworms, millipedes, symphylids, and the larvae of cabbage root fly, bean seed fly, carrot and mangold fly. They were also used in seed dressings against wheat bulb fly. The government-sponsored *Review of Persistent Organochlorine Pesticides* in 1964 recommended withdrawal or restriction of some of these uses "as soon as practicable", but the use of DDT was not restricted until the Further Review in 1969. Then, widespread restrictions were imposed on DDT since it could be replaced by alternative insecticides, especially in gardens where "the home gardener would not be put to a serious disadvantage".

It is inevitable that such potent and persistent pesticides should produce side effects on soil organisms, but the ecological consequences are often complicated. The chemicals themselves react differently in different soils, soil organisms vary greatly in sensitivity, and a response by some may trigger reactions in others. For example, ground-living predatory mites, beetles and harvest spiders are often more affected by persistent pesticides than their prey. Sometimes this is because of greater intrinsic sensitivity but sometimes a result of their greater activity which renders them more likely to encounter pesticide residues. In either case, the result is often an upsurge in numbers of the prey species.

This is normally only noticed when the prey species is or becomes an economic pest, like the cabbage root fly *Delia radicans* or cabbage white butterfly *Pieris rapae*. T.H.Coaker found that the ground beetles *Bembidion lampros* and *Trechus quadristriatus* and rove beetles of the genus *Aleochara* normally consume a significant proportion of the eggs and young larvae of the cabbage root fly before the latter inflict much damage on crops. However, these beetles were affected by residues of aldrin/dieldrin at concentrations of less than one part in ten million in the soil, resulting from sprays or dusts applied in previous years. Unless there was a high concentration of insecticide immediately around the plants, the cabbage root fly larvae were unaffected, and, with less predation, their better-than-normal survival resulted in yield reductions of up to 70% in cauliflowers, cabbage and brussels sprouts. The repercussions did not stop there; increased survival of the fly in the presence of low pesticide residues was thought to have been a major factor in the rapid increase in resistant root-fly populations which occurred at that time.

A similar story was unravelled in the case of the cabbage white. In this case, DDT gave very good control of caterpillars on the plant at the time of spraying. Control was short lived, however, since new leaves were produced and fresh eggs were laid on these. J.P.Dempster studied the survival of caterpillars in the second generation for three years, and found that there was a marked improvement in their survival as compared with those in unsprayed plots. He showed that this was caused by decreased predation by ground-living predators which normally climbed the cabbage plants and consumed the young larvae. The most important predators in this instance were the ground beetle *Harpalus rufipes* and the harvest spider *Phalangium opilio* both of which were seriously affected by DDT residues in the soil.

Occasionally, sublethal levels of DDT appear to increase the numbers of ground beetles when these are assessed by catches in pitfall traps (jars sunk in the ground). But this technique is dependent on the mobility of the beetles and so a spurious increase in numbers may be due to pesticide-induced hyperactivity. Such abnormal behaviour is likely to diminish a beetle's chance of survival by attracting the attention of birds such as skylarks. A diet of affected beetles, in turn, increases the risk of secondary poisoning, and skylarks have been found to suffer from accumulated doses of pesticide, like the American robins mentioned earlier.

Chlordane has been used to kill earthworms in sports turf; other organochlorine insecticides are absorbed and accumulated by worms with no apparent ill effects. The uptake of pesticides depends on the feeding behaviour of the worms and on where pesticide residues are concentrated. *Lumbricus terrestris* can accumulate large amounts of DDT from fallen leaves in sprayed orchards (It was mainly this species on which the American robins fed). It acquires less pesticide from residues in the soil than do shallow-living species which ingest soil. Usually, we know too little about significant physiological and behavioural differences among soil invertebrates to understand how they respond to pesticides. The problem with the *Arion hortensis* group of slugs was mentioned earlier; pesticides will remain a blunt instrument so long as differences in ecology are overlooked.

Most organochlorine insecticides have now been superceded by organophosphorus compounds, carbamates and others, with names like fonofos, chlorfenvinphos, bendiocarb and carbofuran. Sometimes the names suggest their chemical affinity but usually they are simply trade names. The 1988 *UK Pesticide Guide* gives 19 organophosphorus and carbamate insecticides that are used to control soil pests, whereas a little over 20 years ago there was only parathion. Compared with organochlorine compounds, these substances are much less persistent in soils and in the bodies of animals absorbing them. They still kill non-target organisms and cause imbalances in the fauna, but the effects are not as long lasting and they are less readily passed on to predators.

Table 10 Total annual usage of pesticides 1980-83 in agriculture and horticulture. (From MAFF Pesticide Usage Survey Report No 41) including fruit, vegetables and grass..

	tonnes active ingredient	*treated ha*
insecticides, acaricides, molluscicides	1,394	2,311,872
seed treatments	300	3,883,356
fungicides	4,341	6,715,230
herbicides	26,360	12,402,256
growth regulators, sterilants etc	3,138	801,197
TOTALS	35,536	26,114,203

Table 11 Pesticide usage on major crops UK 1986/87 (From BAA Annual Report 1988). Some crops are treated several times a year, eg fungicides on cereals and potatoes.

	treated areas (thousand ha)				
	cereals	*potatoes*	*sugar-beet*	*peas*	*oilseed rape*
insecticides	875	94	139	153	321
fungicides	9002	613	–	–	466
herbicides	7858	182	913	263	723
growth regulators	2120	–	–	–	–
TOTAL AREA GROWN	3922	148	201	158	421

Carbaryl and phorate are poisonous to worms but, surprisingly, the real danger to worms has been from fungicides. Formerly, copper-based sprays, such as Bordeaux mixture, nearly eliminated worms from some orchards, and prevented normal incorporation of leaf litter into the soil as described in chapter 5. The present danger is from the large-scale use of benomyl, carbendazim or thiophanate-methyl on cereals. Since 1985, over half the winter wheat and winter barley crops in Britain have been sprayed with these compounds. They are highly toxic to worms whether absorbed directly from soil or from decaying straw some months later.

New pesticides are continually being developed and there is a need therefore to monitor their effects. A growing number of people, albeit still a minority, buy food grown without using pesticides. This is an option in the western world where food surpluses are produced; in Denmark, it is a thriving industry. Organic farming is considered at the end of this chapter.

Herbicides are used in larger quantities than all other pesticides put together. Table 10 shows that during 1980-83 some 26,360 tonnes per year of herbicides were applied to about 12.4 million hectares in England and Wales. They seldom affect the soil fauna seriously, even after repeated applications. The main effect of total weedkillers such as chlorthiamid, paraquat, glyphosate, simazine and other triazine compounds is to remove the protective plant cover and food source. Paraquat and diquat are exceptional herbicidal substances since they kill green foliage on contact but are then rendered harmless by being adsorbed irreversibly on clay minerals and organic matter in the soil. This property was a great advance because it allowed direct drilling of a new crop within hours of spraying the surface vegetation of weeds and volunteer cereals.

This account of pesticides would be incomplete without mentioning soil sterilants. These are used for the control of diseases, weeds and especially nematode pests, like the potato or cereal root eelworms (see chapter 5) which are otherwise difficult to control. Volatile liquids include carbon disulphide, methyl bromide, chloropicrin, formaldehyde and 'D-D' (a mixture of dichloropropane, dichloropropene and some propane). In addition, there are some special chemicals which decompose and release an active fumigant when applied to moist soil. Examples include nabam, which decomposes to liberate carbon disulphide, and dazomet, which decomposes to form methyl isothionate. Such compounds kill harmless species as well, and although the chemicals may disappear in a few weeks, it may take a year or two for the fauna to recover. Fumigants are generally used in glasshouses or nurseries so their effects are localized.

An unusual example of soil sterilization arose a few years ago when the Ministry of Defence treated part of Gruinard Island off the north west coast of Scotland. The island had been used for experiments in 1942–43 on the release of anthrax spores *Bacillus anthracis* which might be used in biological warfare. The resulting contamination of the soil by these long-lived and virulent spores had prevented any subsequent habitation by man or live-stock. However, after about 40 years the affected area was considered small enough to undertake decontamination. After some initial trials with several disinfectants, 5 per cent formaldehyde in sea-water was chosen and used successfully in 1986 on three hectares of blanket bog. Samples taken afterwards showed that all spores in the topsoil had been killed. The vegetation was almost totally killed as well and has recovered only slowly because the use of fertilizers has attracted heavy grazing by rabbits.

Breakdown of pesticides

The eventual removal of herbicides from soil is, of course, a vital prerequisite for the continued use of soil for growing plants. The same applies to the residues of any other potent synthetic pesticides. Some of these chemicals, including formaldehyde, are removed or degraded by various processes such as volatilization, leaching by rain, photochemical oxidation, hydrolysis and other chemical means, but most important of all are the activities of the soil microorganisms. Fortunately, soils harbour a most abundant microflora and fauna. Although their numbers and diversity may fluctuate in response to constantly varying conditions, there is a continuing turnover of organic material involving the breakdown of many complex substances including synthetic compounds applied by man.

The first synthetic weedkillers were the phenoxyacetic acid compounds 2,4-D and MCPA, known as growth hormone herbicides from their mode of action. Even before their practical use, the possibility of their decomposition by microorganisms had already been recognized. The half lives of these two substances in soil rarely exceed four to six weeks under normal conditions and with recommended dosage. The related herbicide 2,4,5-T is much more stable; it is a useful agent for dealing with woody plants, but its persistence in soils is a disadvantage. (It is no longer used in Britain because of the danger to spray operators from impurities which could be present)

The microbial breakdown of the phenoxyacetic acid herbicides has been the most investigated, and a number of bacterial species and some fungi have been shown to be responsible. The interactions between soil microorganisms and these herbicides are interesting because they illustrate the effects of different factors on pesticide breakdown in soils. When 2,4-D, for example, is sprayed on the soil for the first time, there is generally a lag during which the concentration of herbicide in the soil solution remains more or less unchanged. This period may last for a few weeks, depending on temperature and soil conditions. Then follows a shorter period, perhaps a few days, during which the herbicide concentration steadily declines until little if any remains. During the lag period it is generally accepted that potential microbial decomposers are in some way adapting themselves, i.e. developing suitable enzyme systems, to metabolize the unfamiliar chemical (see p. 125). Such adapted microorganisms can use the herbicide molecules as nutrients and so have an advantage over other organisms. A population of adapted organisms is then produced corresponding to the period of herbicide decline.

Addition of more herbicide allows further growth of the population of adapted organisms without any lag period until all herbicide is consumed. Without any further addition of herbicide, the 2,4-D-decomposing microbial population can remain in the soil for a long time, declining only very slowly. When grass turf was treated monthly with large doses of 2,4-D for two years, one of us (N.W.) detected the presence of 2,4-D-decomposing bacteria more than a year later after 2,4-D treatment was discontinued. Under such conditions, repeated herbicide treatment with 2,4-D becomes much less effective, and it is advisable to use a different herbicide for continued treatment.

Enzyme adaptations by bacteria or other microorganisms are usually specific to one chemical compound. Occasionally, the specificity is not absolute and cross adaptations may occur. Thus the two herbicides MCPA and 2,4-D have something in common in their chemical structure, and microorganisms that attack MCPA are able to degrade 2,4-D also. However, the converse does not apply, since treatment of soil with 2,4-D does not accelerate subsequent breakdown of MCPA.

Adapted microbes can utilize these herbicides as a source of food and energy and achieve complete breakdown of the substance to carbon dioxide and water and ionic chlorine. Many other synthetic chemicals are not completely metabolized in this way but may be partially degraded by the process known as co-oxidation or co-metabolism. Then a succession of different microorganisms may be necessary to effect complete decomposition of the compound.

Microorganisms are largely responsible for the breakdown or detoxification of insecticides and fungicides, and their interactions are of great interest. For example, the organophosphorus insecticides owe their activity to interference with the acetyl cholinesterase enzyme of insects. It is fortunate, therefore, that some soil bacteria possess *constitutive* 'phosphatase' enzymes which, in a few cases, can hydrolyse the phosphate ester grouping in the insecticide molecule, and so inactivate the insecticide completely. The very toxic parathion can be inactivated by several soil bacteria in this way.

As with herbicides, repeated additions of an insecticide to soil can induce the growth of a population of bacteria able to decompose it. An example is the use of diazinon to control the lettuce root aphid. The failure of control after repeated use of diazinon was shown to be due to a population of a *Flavobacterium* species that could decompose diazinon. A similar phenomenon was seen in the case of carbofuran that was used repeatedly in soils to control a rice pest. The effectiveness of the insecticide declined after repeated use.

A vast amount of information is now available on the behaviour and breakdown of pesticides in soil, and repeatedly the activities of microorganisms have been shown to be vital to maintaining soils in a healthy, fertile condition. There still remain certain very stable chemicals which resist biological attack. These are referred to as recalcitrant substances and include the persistent insecticides and herbicides already mentioned, for example DDT and 2,4,5-T.

Leaching of nitrates

In addition to environmental problems caused by stubble burning and pesticides, agriculture is partly responsible for a third issue of environmental concern, that of nitrates in water. From the soil point of view, it is simply wasteful to pour valuable nutrients down the drain, and reprehensible if they cause undesirable effects subsequently. We have mentioned already the ease with which nitrates may be leached down the soil profile, beyond the reach of crop roots. Where there are field drains, water percolating through the soil will mostly be intercepted and flow out into ditches. This gives a means of measuring losses by inserting devices such as a V-notch weir and flow meter, and sampling the run-off at intervals.

Early measurements were made of leaching on the Broadbalk field at Rothamsted during 1878–81. Because of the planned lay-out on Broadbalk, it was possible to compare losses of nitrate arising from different fertilizer treatments – differences between manures and inorganic fertilizers, between ammonium-N and nitrate-N fertilizers, and between spring and autumn applications. The results showed that the time of year was the most important factor; nearly three times as much nitrogen was lost when fertilizers were applied in autumn because the wheat could not take up the nitrogen before it was leached by winter rains. Farmyard manure was also applied in the autumn but the nitrogen was released more slowly and so relatively little was lost.

Similar studies were carried out in County Antrim, N. Ireland, a century later. Here, the experimental site was in an area of intensively managed grassland receiving heavy doses of nitrogen – about 300 kg/ha in artificial fertilizers in three or four doses annually, together with 60,000 litres of pig slurry containing a further 100 kg N. The grass was grazed at intervals by cattle or sheep, and two or three cuts were taken for silage each year. Nitrates lost by leaching were measured daily for 170 weeks, from January 1981 to March 1984.

This intensive monitoring programme showed that nitrogen losses were correlated with the flow of water through the drain; increased flow led to an increase in N collected, with the greatest losses after heavy rains following long dry periods. In 1982, for instance, 35.5 kg/ha was lost by leaching over the whole year, but 21per cent of this was flushed out by two days of heavy rain when 20mm fell in late September after nearly a month of 'Indian summer' weather. Again, after a very dry summer in 1983, heavy rains in December resulted in large losses of nitrogen, reaching nearly 9 kg/ha in a week. On the other hand, much less of the applied nitrogen was leached out in 1981 which had a wet summer.

These results were not just a local phenomenon. Exceptionally high nitrate peaks were recorded in rivers draining arable catchments in England after the 1976 summer which was the driest on record in much of Britain. It seems that a hot, dry summer favours nitrification of organic reserves in the soil, and free nitrates then accumulate because plants cannot fully assimilate them during dry conditions. With the coming of autumn, growth is held in check by the colder weather so heavy rain then flushes out the nitrates before they can be used. In a wet summer, some of the nitrates are lost by leaching but much more are taken up by the active growth of plants.

One must put these nitrogen losses in perspective. Losses of gaseous ammonia from pig slurry applied to the soil in the Irish example could well amount to 40–60 kg/ha N per year, considerably more than by leaching. (Gaseous losses could contribute to 'acid rain' but not to water pollution). Loss of phosphorus by leaching was only about 1–2 kg/ha per year; this was less than that applied and so would indicate an annual net enrichment of soil phosphorus levels, as has occurred in the UK generally since the 1940s.

In Nature, alternate drying and re-wetting of soil often takes place with the consequence that part of the living biomass may be converted into dead organic matter. The latter is consumed or broken down in stages by microorganisms until it becomes finally mineralized to inorganic constituents which serve as plant nutrients. After each cycle of drying and re-wetting, there is a flush of microbial activity in the soil and concomitant mineralization of the soil's biomass. Provided that other conditions are favourable, plant growth is naturally stimulated by this provision of nutrient material. This phenomenon has been frequently observed, especially in tropical areas where there are characteristic wet and dry seasons. In some parts of Africa, for example, crops may be sown just prior to or at the beginning of the rainy season in order to exploit the flush of nutrients that are released when the soil is rewetted. If crops are sown in good time, the leaching and loss of nitrates can be avoided. Leaching implies movement down the soil profile, but in some areas where soils are subject to intensive drying, the upward movement of soil water due to capillary action can transport the dissolved nitrates to the surface. After evaporation on the surface, deposits of nitrates can occur.

Returning to the British scene, several thousand kg/ha of organic nitrogen in old grassland are potentially mineralizable and then subject to leaching. In one measured example, the ploughing of virgin grassland led to the loss of more than 200 kg/ha N by leaching during the first winter, compared with less than 5 kg/ha lost from a similar area remaining under grass. Such nitrate losses are more insidious and less easily measured when they occur from naturally well drained soils overlying chalk or limestone. Here, dissolved nitrates filter down gradually at a rate of a metre or so a year until they reach the underground water reservoirs. By taking measurements at various depths and times, it is possible to calculate when a particular band of nitrate left the soil, and how long it will take to reach the aquifer. Conversely, by running the equation backwards, one can calculate when it left the surface. Much of the present nitrate pollution of ground water is thought to date back to the wide-scale ploughing up of grassland during and after World War II.

Water drains more quickly when the rock is fissured; thus nitrate levels are much higher in aquifer water underlying the Lincolnshire Limestone, for instance, than in most chalk aquifers. However, it is predicted that even the levels in the limestone waters will rise further as stored nitrate is flushed from the finely porous stone.

Fig. 63 Gully erosion in a field sown with oil-seed rape, Penrith. (Photograph by R. Evans.)

Erosion

In 1982, representatives from eight EEC countries met in Florence to discuss soil erosion and conservation. In surveying the problem, D. W. Sanders of the Food and Agriculture Organisation stated that "large areas of the world's agricultural land are now suffering from severe erosion and related forms of land degradation. It is estimated that between 5 and 7 million hectares of land are lost annually through soil degradation. This must be stopped if the world is to retain the potential to feed future populations."

The figures may change but the message has remained the same since G. V. Jacks and R. O. Whyte wrote *The Rape of the Earth* in 1939. We are used to hearing about soil erosion in the Third World and in the U.S.A. but, unless we have seen an affected area, dry statistics do not evoke strong images or emotions. In Britain, we tend to think of erosion as something affecting Cat Bells in the Lake District, or Mam Tor in the Peak District, or the Three Peaks area in the Yorkshire Dales, where excessive trampling has worn away the vegetation cover and exposed the soil to gully erosion by heavy rain. It was a surprise, therefore, to many people when the Soil Survey published its warning that agricultural land in Britain was suffering serious erosion, and set up a 5-year programme in 1982 to monitor it. A series of air photographs in fifteen areas was used to identify rills and soil outwash at the bottom of slopes. Field surveyors then visited sample sites to check the photographic interpretations, and to measure the amounts of soil deposited.

Water erosion of farm land in this country is not new, for Whyte mentions bald areas in fields, piles of eroded soil on roadways, and the value of hedges and contour trenches in arresting soil movement in hilly districts. However,

Fig. 64 Deposition of soil from water erosion in Cambridgeshire. (Photograph by R. Evans.)

it has become significantly worse in the last twenty years as a result of farming innovations: first, the ploughing up of old grassland, especially the chalk downland slopes and previously uncultivated parts of Brecon, Radnor and Exmoor; secondly, the increase in field size due to removal of hedges, which had often been placed at critical changes in slope, and had served to check the downhill drift of soil; and thirdly, the introduction of continuous arable cropping, and the use of tramlines which, because of their compaction, inhibit infiltration and start run-off.

Water erosion is a fairly local problem at present, affecting mainly light and coarse textured soils on valley sides, but about one fifth of cultivated land in lowland England and Wales is potentially at risk (Figs 63 and 64). The effects are mainly seen in an insidious loss of topsoil which causes a slow deterioration of soil potential and crop yields. Of equal, or in some cases, greater concern are the problems produced by increased sediments in streams and rivers. Increased turbidity and silting reduce the quality of trout and salmon rivers, and can also lead to flooding.

Control measures are generally only taken when silting results in blocked drains and flooding. Apart from returning the arable land to grass or replacing contour hedges, the only ways of limiting erosion on sloping ground are by reducing run-off over bare soil. This can be done by early drilling, by cultivating fields on the contour rather than up and down, and by using tines behind a tractor to break up compacted tramlines, so that water can penetrate into the soil instead of flowing over the surface.

Erosion of upland grassland and moorland has also become a problem in some areas, such as the Dark Peak in Derbyshire, around Plynlimon in Wales, Grasmere in the Lake District and in Teesdale in North Yorkshire. Here, the cause is the increase in the numbers of sheep since the 1930s, which has been encouraged, especially since the war, by agricultural policies. The number of sheep in the Peak District has increased three-fold during this period. The deleterious effects of overgrazing were noted in the Draft Structure Plan for the Peak Park in 1974 and this led to a Moorland Erosion Study. In an analysis of the problem, R. Evans considered that erosion on steep grassy slopes was initiated by grazing at a density of about two sheep per hectare, and probably by less than this above 435 metres altitude. On the flank of Black Tor in the Derwent catchment area of the Dark Peak, a series of scars was monitored between 1975 and 1976 and showed an increase in bare soil of 42 per cent. Sheep scars are now obvious on many steep slopes in the U.K.

Water and wind erosion present the obverse and inverse faces of a coin so far as climate and topography are concerned. Wind erosion is important only in the drier eastern counties on flat, open expanses of intensive arable agriculture. The soils most at risk are the light sandy soils and peaty fens of East Anglia, Humberside and the Vale of York. Here the danger periods are during dry springs, when winds of 15 mph or more blow for several days at a time over fields with little or no crop cover.

One of the worst fenland blows in living memory occurred in 1968 – in the same year and area that the disastrous July floods occurred which were mentioned earlier (page 139). E. Pollard and A. Millar, from the Monks Wood Experimental Station, witnessed the scene a few miles away and recorded the events. Rainfall in February of that year was about 60 per cent of the average for the area, and March had less than half the average. On the 16th of March, the wind began to blow and continued for the next thirteen hours at speeds of 22 mph or more, which is considered a strong breeze, force 6 on the Beaufort scale. Drying winds blew during the following four days at about the same speed but with gale force gusts of 44 mph on March 18th and 20th. "The results of this combination of weather conditions were quite spectacular"; virtually every field that had been finely cultivated for growing sugar-beet around Holme and Ramsey St. Mary was on the move. Particles of black peat soil were visible up to a height of 60 m, and where the wind was channelled by buildings or shelter belts, visibility was reduced to 10–20 m. Apart from the direct loss of precious topsoil, often with seeds and fertilizer, there was also the cost of digging out blocked ditches which amounted altogether to £100 per hectare, according to local estimates.

The measures that can be taken to prevent or minimize erosion are mainly good husbandry with some help from modern technology. Conspicuous losses due to wind erosion, such as those described above, have led to a variety of cultural practices for stabilizing the soil surface and reducing wind speeds at ground level. Where the peat has already worn thin, deep ploughing can bring up clay which, like marling, is most effective in holding down the soil. Elsewhere nurse crops, such as spring barley, can be sown, or lines of straw planted at right angles to the prevailing wind, to provide temporary wind resistance and shelter. The nurse crop is later removed with a selective herbicide such as 'Fusillade' (fluazifop-p-butyl), a light dose sometimes being first used to check its growth before a second dose administers the *coup de grace*.

In sandy districts, the simplest technique consists of cultivating during wet conditions to create a rough, cloddy surface, into which the crop is drilled.

Nowadays, a crenelated surface can be produced with a 'furrow press' plough, the sugar beet being sown across the grain. A synthetic latex resin, such as vinyl acetate, is sometimes sprayed on the surface in bands to protect high quality crops. This technique was used in the early 1960s to stabilize sand dunes on the Norfolk coast, after they had been breached by storms. The emulsion bound the sand together until marram grass had grown through again. The agricultural uses have been developed at the Gleadthorpe Experimental Husbandry Farm in Nottinghamshire.

Organic farming

It is not possible in this book to do justice to the topic of organic farming but it deserves mention as it has a strong following and implications for soils. Modern intensive agriculture now needs great inputs of nutrients to sustain soil productivity, and an elaborate spraying programme for weed, pest and disease control to realize it. Proponents of biological husbandry believe that such high technology agriculture cannot be maintained indefinitely nor, indeed, is it always appropriate. They consider that much more use could be made of natural biological processes that are neglected in conventional modern farming. The controversy has run for at least 25 years, and it is interesting to compare the changes that have occurred within the movement since J.Jenks wrote *The Stuff Man's Made Of* in 1959. Like Howard, quoted in the previous chapter, Jenks believed that the soil held a key position in the wholeness of nature, and that all non-living amendments, such as man-made fertilizers and pesticides, were fraught with danger; their increasing use could upset 'the balance of nature' so that civilization itself would founder.

Such extreme views prejudice their own case if they are not quickly borne out by experience. Nevertheless, echoes of this fear return from time to time, as in the terms of reference of the Agricultural Advisory Council in 1969 which were to consider whether "the inherent fertility of the soil was being eroded and the fundamental structure of the soil damaged beyond repair". The 'orthodox' school can also be accused of adopting a dogmatic view at times. A German programme of agriculture and forestry research stated (in translation) that "these experiments and investigations are most important in order once and for all to get a precise basis for a final rejection of the erroneous teaching of the enemies of our ways of fertilizing".

The 'muck and magic' school, as it was once called, has survived and gained in standing by adopting a more critical approach and dropping some of its wilder claims. The current attitude is well summarized in a recent book *Biological Husbandry: a Scientific Approach to Organic Farming*. Here we read that the movement embraces two lines of approach. First, there will always be a purist school. This eschews the use of all artificial agrochemicals which have come to dominate so-called conventional agriculture in the industrialized countries. It relies on sound crop rotation practices with a careful balance between crops and stock, or between crops, green manures and legumes. Timeliness of operations, such as ploughing and cultivating, is also paramount. This can only be achieved on a farm scale with the proper equipment – no room for your 'good life' amateurism here. Such an approach may be thought of as providing a standard, or control, against which more conventional practices can be compared.

Secondly, there will also be those who believe in a "complete biological approach...but with the use of chemicals for problem solving or crop topping-up, which will not do any harm to the biological life or condition of the soil". The most orthodox farmer or soil scientist could say amen to that – if it did not cost too much. Today, much research is directed towards monitoring pests such as aphids and pea moth so as to avoid unnecessary, prophylactic or insurance spraying. On a global scale, the more 'purist' organic farming approach is likely to cause more famine than 'chemical' farming. The recycling of nutrients needs to be combined with the selective use of agrochemicals to control difficult pests and diseases.

Recent measurements in Britain of net farm income, per £ of energy used, suggest that organically run farms are not markedly different from conventional ones. This conclusion was based on a survey of 70 organic farms and 40 conventional ones by a research team from the Department of Economics, University College of Wales at Bangor. Although the majority of organic farmers had lower financial returns, some outperformed the conventional average. Furthermore, organically grown produce can command a higher premium which can shift poor returns to good ones. This higher premium depends, of course, on a favourable supply:demand factor; if more farmers entered this market the premiums would be reduced.

No one questions the value of organic matter in the soil for making cultivations easier and improving seed beds. However, organic farming by itself does not increase soil organic matter unless very large quantities of farmyard manure are used, or a system of ley/arable farming is adopted. Measurements made by the Agricultural Development and Advisory Service at the Soil Association's farm at Haughley, Suffolk, did not show extra total soil organic matter compared with adjacent fields receiving conventional fertilizers and management.

As a source of major nutrients, the main difference between manures and inorganic fertilizers is in the slower rate at which nitrate-N is made available to crops. This may be an advantage as less may be lost by leaching, as shown in the Broadbalk experiment described earlier. Because of its high water content, however, manures cost more to transport and spread: the NPK in a tonne of farmyard manure can be matched by 37 kg of fertilizer, 1⁄27th the weight. On a national scale, there would also be a very real problem of retrieving and recycling the nutrients carried from the farms into the cities. This would be extremely expensive but the time may come when society will have to consider seriously the pollution and waste of resources entailed in the present policy of sewage disposal.

In practice, it is difficult to draw a clear line between organic husbandry and conventional farming. It is an attitude of mind as much as anything. Some would consider it a test of man's stewardship of the Earth. An American survey showed that farms run on organic lines were also most energy conscious and employed methods designed to minimize water pollution and soil erosion.

We have yet to see whether British farmers will reduce their use of agrochemicals on crops in response to pressures to reduce productivity, or will maintain intensive use on reduced areas. The Set-Aside scheme does not allow grass strips around arable fields to be used by livestock, so there is no encouragement for a return to more mixed farming.

10

Reclamation and Restoration

In this country, we do not have the massive problems of soil reclamation which face many parts of the world. Our reclamation problems arise not from deforestation, prolonged overgrazing, or the tillage of fragile soils subject to drought or erosive monsoon rains, but from the mining of mineral resources and the dumping of industrial wastes. We have, in the past, created enormous areas of derelict land from the mining of coal and oil shale, clay and gravel, chalk and limestone, ironstone and fluorspar. In such a crowded country, this has become quite unacceptable. An outcry against the despoliation of the landscape reached a pitch in the 1960s with such books as John Barr's *Derelict Britain*. As a result of government legislation, there has been a concerted effort since then to reclaim some of the worst areas, and to ensure, in future, that restoration of the land is planned, as far as possible, as an integral part of any new mineral working. Today, it is widely recognized that good restoration is the key to future planning consents.

However, even with the best programme of strip mining, in which subsoil and topsoil are replaced, major physical and biological disturbance is inevitable. In more severe cases, the topsoil may become degraded through prolonged storage while, in the worst cases, materials may be exposed or tipped which have few or none of the characteristics needed to support life. Yet all natural soils are derived from the weathering of initially inhospitable substrates; there must be few that can outdo the products of Nature's volcanic furnaces or glacial epochs. The spoil from the Northamptonshire ironstone workings is, after all, little different from the boulder clay that was left when the glaciers retreated. This chapter looks first, therefore, at some of Nature's solutions to the problems of building an ecosystem on such materials, and then examines some of the human solutions for restoring or re-creating soils with as many desirable features as possible.

Natural processes

Some aspects of soil weathering and maturation depend on very slow processes as described in chapter 2; they may require centuries or longer periods of time. However, plants are themselves agents in initial soil formation. The sequence of natural vegetation development seen in disused mineral workings gives us clues to the development of nutrient cycling, and thus to an unseen world of microorganisms at work in the nascent soil.

A good example of natural colonization was seen at Clipsham Quarry in Leicestershire. Here, after about 40 years, the centre of the old quarry floor had from 5 to 12 cm depth of friable 'soil' overlying the bedrock. It contained very low amounts of organic matter and plant nutrients – about 10 per cent as much total nitrogen, 8 per cent as much 'available' phosphorus, and 17 per cent as much potasium as a near-by old pasture, itself rather low in

Fig. 65 Quarrying exposes the parent rock. If the quarry is then disused, a new skeletal soil slowly forms by natural processes of weathering and plant colonization. Ardley Quarry, Oxfordshire (1975). (Photograph B.N.K.D.)

nutrients compared with most arable soils. The average yearly rainfall was also small, around 600mm, so water deficit was another limitation on plant establishment. In 1980 the general appearance was of a stony, almost barren area of ground, but closer examination showed that it supported 7 species of mosses and 46 species of flowering plants, such as eyebright, wild strawberry, marjoram and bird's-foot trefoil, occupying about 8 per cent of the surface. The biomass below the surface was probably greater than above ground, for many plants had extensive root systems seeking water and nutrients.

This proto-soil had also become colonized by five kinds of earthworms, notably by *Lumbricus terrestris* whose presence was often betrayed by small aggregations of stones that had been dragged towards its burrow (see chapter 5). Their biomass was not inconsiderable, and they would be important agents in the incorporation of dead vegetation below ground. A few small snails and some hardy kinds of soil arthropods were also present, especially the drought-resistant oribatid mite *Humerobates rostrolamellatus* and the pill woodlouse *Armadillidium vulgare*, contributing to the breakdown and cycling of plant material.

Older parts of this quarry, and other chalk and limestone quarries, show how such pioneer communities can help to develop a closed grassland turf not unlike that found on thin Rendzina soils of the chalk downs or upland limestone pastures. Some quarries become colonized by shrubs and trees whose roots penetrate more deeply, fissuring the rock, bringing up nutrients, and enriching the lower soil horizons with organic matter when they die. As vegetation covers the ground, it acts as a protective blanket, reducing the destructive effects of wind and rain as well as the extremes of temperature to which the bare surface was exposed (Fig. 65).

Increasingly, today, such old quarries are either reworked or used as receptacles for spoil or domestic waste. The problem of restoring vegetation on such landfill sites is dealt with later. It is worth noting, though, that certain quarries are important outdoor museums for their geological exposures, while some naturally colonized quarries are interesting and attractive places in their own right simply because of their immature ecosystems.

Many kinds of plants may participate in this colonization process in chalk and limestone quarries because, although there is a deficiency of nitrogen, phosphorus and potassium, there is often an abundance of magnesium, iron and, above all, calcium in the rock. Where calcium and magnesium are also lacking in the parent rock, as in granite, vegetation has a still greater struggle to become established. This situation occurs in the abandoned waste tips of the china clay industry in Cornwall, for these are derived from the granite cores of ancient volcanoes. China clay, or kaolinite, is itself an extremely pure and fine material (an aluminium silicate, see chapter 1). It is washed out from the decomposed granite leaving vast quantities of quartz particles, with a certain amount of mica and clay, in the older workings. This waste material cannot be put back into the holes, for they are virtually bottomless as far as the industry is concerned, so it is heaped into giant sand castles.

Not only do these wastes contain extremely small amounts of all the major plant nutrients, but they often lack even the basic minerals from which such nutrients could be derived. Natural plant colonization on these tips is therefore a very slow process, and is closely linked to the accumulation of nutrients from the air. The spoil receives nitrogen, phosphorus, potassium, calcium and magnesium from atmospheric depositions – dust and rain – but these are also rapidly leached away so the net accumulation rates are very low, and perhaps even negative in the case of potassium. Colonization seems to start 25–55 years after tipping has ceased, and consists at first of heather, gorse and broom with a few other plants. This community is later invaded by goat willow *Salix caprea*, followed by woodland vegetation with rhododendron, birch and oak when the tips are 75–120 years old.

Once the vegetation is established, nitrogen actually accumulates faster than would be expected from atmospheric inputs alone. This is due to nitrogen fixation by the root nodules of gorse and broom (see chapter 3); gorse can fix 26 kg nitrogen per hectare per year, compared with a measured atmospheric deposition of only 9 kg/ha/year. At a few sites, tree lupin *Lupinus arboreus* is the first colonist when the tips are only 10–20 years old. This is not a native British species but it has become naturalized in some areas, and is an important plant on these sand wastes because it is able to fix relatively large amounts of nitrogen – about 72 kg/ha/year. Individual plants die after about six years, but the boost they give to the nitrogen economy of the developing soil is sufficient for other species, that do not fix nitrogen, to become established around them. The critical factor, then, for building up organic matter in these wastes, and stimulating plant succession, is the capture and cycling of nitrogen for which woody legumes are the prime agents. It seems that non-leguminous shrubs and trees do not begin to colonize china clay sand

Fig. 66 Plot experiments for establishing vegetation on sand wastes from china clay workings in Cornwall. (Photograph R.H. Marrs.)

wastes successfully until a nitrogen capital of some 1000 kg/ha has been built up, of which about 300 would be in the living plants at any one time, and 700 in the soil. This story seems simple in retrospect, but it involved a team of scientists under the leadership of A. D. Bradshaw to discover the principles and to work out the details. Figure 66 shows some of the experimental plots that were set up.

The mountains of deep-mined shale which used so to dominate parts of the landscape of south Lancashire, Yorkshire, Northumberland and Durham presented a greater range of problems for vegetation establishment than the china clay tips. These heaps could be up to 200 ft (65 m) high with slopes as steep as 35° and littered with many coarse fragments the size of dinner plates. Apart from their instability and tendency to erosion, many of the tips showed persistent toxicity resulting from the extreme acidity of the shale. The pH of some types of shale could be as low as 1.5–2.0 owing to the presence of iron pyrites. Chemically, this mineral is iron sulphide (FeS_2), which was formed under anaerobic conditions in the swamps where the original Carboniferous deposits were laid down. When iron pyrites occurs in large crystals, it has a golden colour from which it has gained the name of 'fool's gold'. More often, however, it occurs in coal shale as tiny strawberry-like granules – framboidal pyrites – in which state it is highly reactive towards

Table 12 Soil profile on a coal shale waste heap after 100 years. (From I.G.Hall 1957)

depth(cm)	*Profile description*
0-8	Dark brown loam, large crumb structure, mully spongy, very porous and friable ... occasional worms ... pH 4.8, fairly distinct from -
8-20	Brown loam, crumb structure deteriorating with depth ... well rooted, worms, merging into –
20-31	Grey silt loam, frequent large shale fragments and abundant small fragments, structureless ... merging into blue grey, scarcely weathered shale.

oxygen and water. A complex sequence of chemical changes is thought to occur in which the net effect can be summarized as follows:-

iron sulphide + oxygen + water yields iron sulphate + sulphuric acid

$$6\ FeS_2 + 21\ O_2 + 6\ H_2O = 6\ FeSO_4 + H_2SO_4$$

An interesting feature of this process is the mediation of a particular bacterium, *Thiobacillus ferrooxidans* (see chapter 6), which appears to be necessary for two of the intermediate stages. Without it, the full oxidation of pyrites does not occur or proceeds much more slowly. If there is much pyrites present, sites can actually deteriorate with time because formation of the acid may continue for a hundred years, or possibly several hundred years, before all of it will have been removed by natural leaching. On the other hand, if there are substantial amounts of carbonate minerals present, such as ankerite or siderite, these can act like lime in neutralizing the acid. High acidity is directly toxic to plants, and it also initiates a series of other unfavourable conditions in the spoil: by lowering the exchange capacity of the material, not only are beneficial ions released and then more readily leached away, but harmful ions, such as aluminium, come into solution and exert phytotoxic effects. The rate of plant colonization, and the kinds of grasses, herbs, shrubs and trees that become established, therefore depend greatly on the net effect of these minerals on the pH of the spoil.

A large population of surface-living springtails can develop quite quickly on some coal wastes, perhaps because of the scarcity of their natural predators. These may perform a valuable role in the cycling of organic matter. The mite fauna tends to be rather simple and is sometimes dominated by one or two oribatid species. *Mixochthonius latipes* is particularly interesting because it occurs where the pH is as low as 2.7, and has been found nowhere else in Britain. If the spoil is limed, numbers of this mite decline in favour of another small oribatid, *Oppiella nova*. Colonies of *M. latipes* presumably occur in some natural pockets of acid soil but these have yet to be discovered. In Denmark and other parts of mainland Europe, this species is associated with beech and oak forests, though even there it was only discovered in about 1950.

Sites that have been left for 50–60 years can remain virtually devoid of earthworms if pH levels remain around 4.5 due to pyrites oxidation. On the other hand, less acid coal shales support reasonable populations of six species after a much shorter period, with the red worm *Lumbricus rubellus* as the most abundant species.

In 1957, I.G.Hall published an account of an extensive survey of over 200

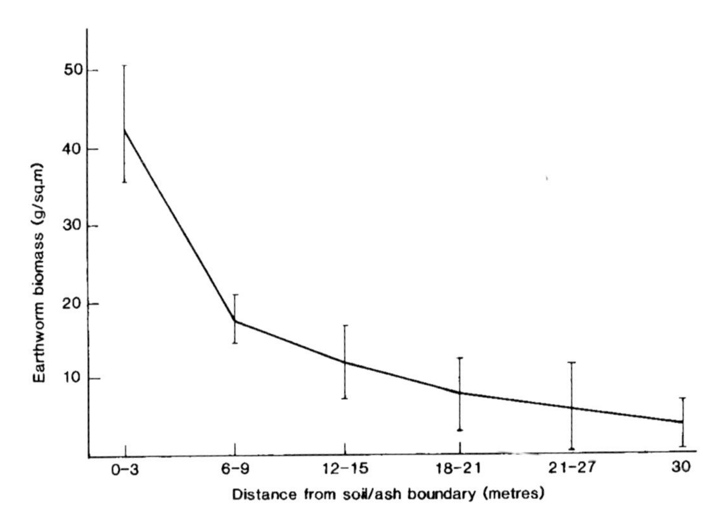

Fig. 67 Earthworm colonization of pulverized fuel ash on a 9 year old site. (From J.E.Satchell & D.A.Stone, 1972.)

disused pit heaps in England in which he described stages in soil formation. After 20 years, the weathering of shale fragments into fine particles extends down to about 30 cm, but not until it is 40–50 years old can the spoil be said to resemble a normal soil, and even after 100 years many spoils lack proper soil structure. Table 12 describes a profile pit from a 100 year old heap in Somerset with young oakwood on it.

Pulverized fuel ash (PFA) is a waste product which remains after coal has been ground to a fine powder and burnt in electricity-generating power stations. The ash consists mainly of spherical glassy particles similar in size to those of a light silty soil, and therefore it has good water-holding properties. Like china clay wastes, it is composed predominantly of silica (SiO_2) but with the addition of considerable amounts of alumina (Al_2O_3) and iron oxides. There are also smaller amounts of the oxides of calcium, magnesium and some other elements, including boron. Nutrient ions for plant growth are therefore present in reasonable quantities – if not immediately available – except for nitrogen and phosphorus which are almost totally lacking. Some plants are sensitive to boron but after about two years of natural weathering the concentration of this element is reduced by leaching to non-toxic levels.

The high salinity of raw PFA is illustrated by the fact that a coastal plant, hastate orache *Atriplex hastata*, is often one of the first colonists. If left to

Table 13 Soil profile on Leblanc process waste after 60-80 years. (From E.F.Greenwood and R.P.Gemmell 1978)

depth(cm)	*pH*	*Profile description*
0-5	7.7	Black, surface humus
5-15	7.7	Dark, partially humified waste
15-35	7.8-8.0	Yellowish-brown waste stained with deposited ferric salts
35-55	9.4-9.7	Yellowish, partially weathered waste. Ferric salts present
55-75	12.2	White, unweathered waste, calcium hydroxide present

Table 14 Some rare plants in South and West Lancashire found on industrial spoil. (From E.F.Greenwood & R.P.Gemmell 1978)

Yellow-wort *Blackstonia perfoliata*
Early marsh-orchid *Dactylorhiza incarnata*
Southern marsh-orchid *Dactylorhiza praetermissa*
Northern marsh-orchid *Dactylorhiza purpurella*
Viper's bugloss *Echium vulgare*
Marsh helleborine *Epipactis palustris*
Fragrant orchid *Gymnadenia conopsea*
Bee orchid *Ophrys apifera*
Green-winged orchid *Orchis morio*
Common broomrape *Orobanche minor*
Round-leaved wintergreen *Pyrola rotundifolia*

Nature, grasses, clovers and other plants soon appear, and eventually build up a mat of roots and undecayed vegetation near the surface. This enables surface-living earthworms to colonize, especially the red worm *Lumbricus rubellus*. Worms cannot survive in the raw PFA because of the high concentration of salts, but as these are leached out, and organic matter is incorporated into the soil from plant roots, burrowing species are able to colonize the PFA. They continue to diversify and build up in numbers until, after about 20 years, a 'mature' worm community with some 15 species is established (Fig. 67). By this time, the waste heap can support a flourishing vegetation with trees and bushes, and one would only be able to identify its industrial origin by digging a soil pit.

Two types of industrial wastes, blast furnace slag and Leblanc wastes, have produced extreme alkaline conditions. Slag from blast furnaces was produced in large quantities by heating iron ores with lime; for every 10 tons of iron produced there were 5–8 tons of alkaline slag which had to be disposed of. Likewise, in the manufacture of sodium carbonate in the nineteenth century by the now obsolete Leblanc process, a slurry of calcium oxide and hydroxide wastes was produced which was run into beds several acres in extent. These beds formed a feature of the landscape in parts of south Lancashire and Cheshire.

Both of these materials in the fresh state have pH levels as high as 12.5, and can scarcely be considered as soils at all. In contrast with the pyritic colliery spoils, however, leaching of these wastes effects a significant change within a relatively short period of time – in this case by reducing their alkalinity. Lichens and mosses are among the first colonists as the superficial layers of these wastes become leached, and thereafter a range of other species with deeper rooting habits gradually become established. After about 40 years, the pH in the top 25 cm of the profile is reduced to 8.0 or less, and after 80 years this degree of leaching extends down to about 55 cm as shown in Table 12. Disturbing a site, therefore, can put the clock back dramatically. Beds that are now 100 years old support some very interesting communities of lime-loving plants including, in some places, vast colonies of 3–4 species of orchids (Table 14 and Plate 13). The oldest, undisturbed sites are becoming colonized by willows and hawthorn which indicate that a substantial nitrogen capital has accumulated, but little is known about the soil fauna and microflora of these artificial soils.

These few examples illustrate several points in the natural development of soil ecosystems on a range of materials disembowelled from the earth or spewed out of the furnaces and vats of industry. The obstacles which have to be overcome for the development of a living soil can be divided into physical and chemical factors. Hostile physical properties range from excessive stoniness, with an absence of fine soil-forming material, to a predominance of clay minerals at the other extreme, as in the Corby ironstone workings, where the main problem is lack of free drainage. Some sites suffer from steep and unstable slopes subject to erosion and slumping – the Aberfan disaster is well known; almost all will experience extremes of surface temperature until a protective vegetation cover has formed. On the chemical side, we can distinguish nutrient deficiency, which is almost universal, from factors that are actually antagonistic to most forms of life, such as the extremes of acidity and alkalinity described.

Practical reclamation schemes must deal with these various problems, and convert the spoil environment into a soil environment within as short a time as possible. Most sites, however, do not pose problems as extreme as those described above. Indeed, site restoration is now usually an integral part of the whole process of mineral working and waste disposal. Modern machinery and technology allows one to 'think big' and not simply to apply cosmetic solutions. The next section describes some examples of soil reclamation with a variety of different starting points and end products.

Soil reclamation

Reclamation, rehabilitation, restoration and re-creation all have slightly different shades of meaning depending on the objectives and the efforts required to achieve them. The key to success in every case lies in the building of a soil ecosystem which can sustain vegetation, and this can seldom be achieved overnight. First, there may be the task of creating a desirable land form using large earth-moving machinery. The removal or reshaping of existing spoil mounds can be a major exercise, but modern operations integrate soil removal with placement as far as possible to minimize effort and costs. Next, there is commonly a need to produce good physical conditions in the top metre or so in order to allow free movement of air and water through the soil. This may again require special machinery for breaking up compacted ground and producing good macro-structure. And thirdly, there are often chemical problems and deficiencies to be overcome, especially when one wishes to restore a site to some productive use.

The simplest objective is the 'greening' of waste heaps or quarry faces to stabilize them or to heal unsightly scars on the landscape. This may involve the sowing of selected grass species or cultivars, the use of fertilizers and/or legumes to raise the nitrogen level in the soil, or planting with particular trees. We can learn a lot from understanding natural processes of vegetation establishment, and then trying to accelerate them by alleviating the constraints. A great deal of research has been directed at rehabilitating the legacy of the past. In modern mineral workings, however, it is generally assumed that the land should be returned to some economic or socially useful purpose: where the land was formerly in agricultural use, for example, planning requirements have usually specified its restoration to agriculture as part of the overall scheme. With the likelihood that much existing agricultural land will come out of production in the future, this goal is less needful. It may be more appropriate to restore areas for public recreation or even to create particular communities for wildlife conservation. Both mineral extraction and agriculture have destroyed many rich wildlife habitats so this last option is attracting increasing attention. This section describes a few examples of these different approaches.

Perhaps the classic example of land restoration, and certainly the most extensive, is that from open-cast coal workings. By 1988, more than 52,000 hectares (27 square miles) had been restored in Britain since the Open-cast Coal Executive was set up in 1942. From the outset it was determined that topsoil should be replaced when the mining was completed. (Open-cast strip mining for ironstone in Northamptonshire had no such requirement.) However, it was quickly discovered that simply levelling the spoil with its mixture of clay, shale and boulders, and covering it with a skim of topsoil, was not enough. Farmers soon complained that they could not grow crops satisfactorily on the restored lands. A new code of practice was therefore introduced which provided for the separate stripping, storing and replacement of subsoil and topsoil to cover the overburden to a depth of 0.9 metres. It also allowed for a five year period of management by the Ministry of Agriculture, involving the installation of permanent under-drainage, before returning the land to the farmer.

Once the topsoil and subsoil are removed from the working area and stored, the excavation proceeds, in the simplest case, rather like a gardener double digging. A dragline opens a cut in the overburden and the exposed coal is removed. Further parallel cuts are then made, the overburden from each new cut being cast into the void of the last one, and the final void backfilled with the overburden from the first cut. Before the subsoil is replaced, the overburden is usually ripped with a special 3-tined, heavy-duty, winged ripper working to a depth of about 0.5 m in two directions at right angles to break up surface pans. Any large stones brought to the surface are also removed. The stored subsoil is then spread in two layers about 0.3 m deep, each layer again being 'rooted' to remove anything that would impede a plough. Finally the topsoil is spread and cultivated. The whole emphasis during this restoration phase is on creating the best physical conditions for life processes to start up again. It is possible actually to improve on the original conditions during this process by inverting strata to bring a lighter material nearer the surface. The main dangers to be avoided are compaction, which would lead to poor drainage, and the loss or deterioration of topsoil.

Soil microbial populations tend to decline in stored heaps but quickly reassert themselves when the topsoil is respread. Aerobic heterotrophic bacteria, *Streptomyces*, ammonium oxidizers and fungi are all active within a few months, and build up to high numbers as they consume dead organic matter. Comparisons in America between a site restored with topsoil and an undisturbed site nearby suggested that the microbial community had begun to stabilize after about four years, as judged by the diversity and density of their soil fungi. Sites without topsoil had a less varied fungal micro-flora, even after eight years, with a great predominance of *Aspergillus* and *Penicillium*, suggesting an immature soil environment.

In the early years of agricultural restoration, the emphasis is on the need to establish good soil structure and to build up the nutrient status. These objectives are usually achieved by sowing a mixture of grasses and clovers with generous applications of fertilizers for several years. A common programme for the first five years might use about 2000 kg/ha of compound fertilizer – about twice the amount normally used on farmland – with 22 tonnes/ha of ground limestone and 500 kg/ha of nitro chalk to counteract any tendency to acidity.

The extraction of alluvial gravel, for example in the Thames valley, creates somewhat different problems. Often the amounts of gravel removed are so large that it is not feasible to fill in the holes that are dug, and so they have been developed as reservoirs and attractive lakes. (The restoration of flooded gravel pits for amenity and wildlife is well developed but does not concern us here.) However, there is also a great need around urban centres for holes in which to deposit the vast quantities of wastes which modern society generates. Furthermore, the gravels sometimes underlie prime agricultural or horticultural land. Increasingly, therefore, there has been an effort to combine gravel extraction, waste disposal and agricultural restoration, using the land three times over, through a rolling programme of extraction, infilling and soil replacement.

If the waste consists of inert materials, such as builders' rubble, the problem is mainly one of careful soil handling with the proper equipment and under good weather conditions. A procedure has been developed by Greenham Sand and Ballast and used with great success for horticultural soils at Shepperton in Middlesex. Essentially, this involves the use of two tracked excavators, for soil removal and reinstatement, and a dumper truck for moving the soil from one site to the other. The topsoil and subsoil are removed and replaced separately in bands a few metres wide. The dumper truck always runs on the exposed gravel or filled surface to prevent compacting the soil, and the excavator at the receptor site also runs only on the filled surface (Fig. 68).

New landfill problems arise when household or industrial wastes are used which contain materials that can readily decompose. In order to understand these problems one needs to know something about the nature of this waste material, its disposal, and the decomposition processes that take place. Nowadays, paper represents about 30 per cent by weight of domestic waste, compared with 24 per cent for vegetable and other putrescible material, 16 per cent for metal and glass, and 14 per cent for dust and ash. Industrial wastes are more varied but contain similar materials. Normally, these components are not separated, and so the heterogeneous mixture has to be crushed and compressed to minimize subsidence later. The waste is built up gradually in layers 2–3 metres deep which are covered by thin layers of inert material such as sand or clay to minimize flies, vermin and wind-blown refuse. For agricultural restoration, the waste should finally be covered with 0.9 m of uncompacted subsoil followed by 0.3 m of topsoil.

Once water gets into the waste, bacteria start to become active and quickly use up the available oxygen so that anaerobic conditions develop. Anaerobic bacteria now take over producing methane and carbon dioxide in roughly equal quantities (55% : 45%) through biodegradation of the paper and other organic materials. Gas production – methanogenesis – usually starts a few months after tipping and may continue for several years depending on the total volume of wastes and how much organic matter drains out as leachate.

Fig. 68 Careful stripping of topsoil and subsoil with excavator and dumper truck before gravel extraction at Shepperton, Middlesex, and their replacement at the receptor site. (Photograph B.N.K.D.)

Although methane and carbon dioxide are not intrinsically toxic to plants, they can effectively deprive the deeper roots of oxygen and thus inhibit growth (Plate 14). Microbial activity also produces heat, and so the temperature in the fill gradually rises, as in a compost heap, to 25–45°C, and sometimes considerably higher. This tends to dry out the overlying soil and may contribute to drought stress in summer.

One way of reducing the effects of methane on the overlying vegetation is to introduce a layer of compacted clay over the refuse as a seal, which will tend to drive the gas out to the boundary. This, too, can have drawbacks as it may affect trees on the boundary. (More serious effects occur if there are nearby buildings into which the gas can migrate). Alternatively, plastic venting pipes can be laid, rather like field drains, which duct the gas passively out to the edges of the field and allow it to disperse into the atmosphere. In some cases it is possible to pump the methane out and flare it off or use it as a fuel for local heating or electricity generation.

Problems associated with reclamation of landfill sites, particularly those concerned with landfill gas emissions, were studied at the Joint Agricultural Restoration Experiment at Bush Farm, Upminster, Essex. This was one of a series of experiments to study the reclamation of sand and gravel workings under the direction of a group of advisers drawn from the Department of Environment, Ministry of Agriculture Fisheries and Food, and Sand and Gravel Association.

The initial objective here was to examine the feasibility of high grade restoration using alternative methods of soil handling. The opportunity was taken to compare the relative advantages of directly transferring soil from one area to another, with the more traditional use of stockpiles. In each case, standard earthscrapers for reinstating soils in the conventional way were also

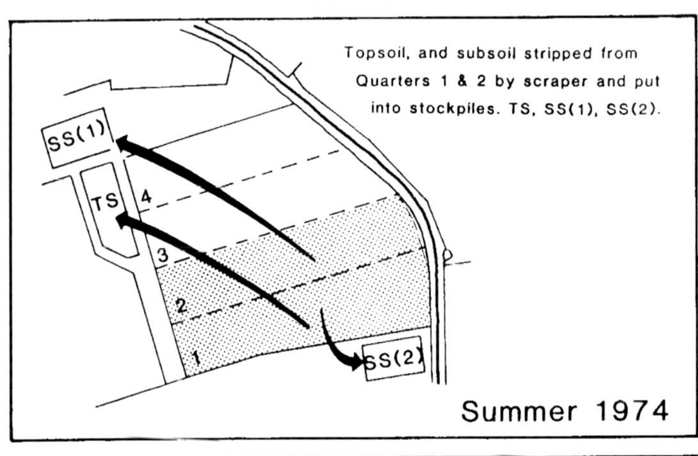

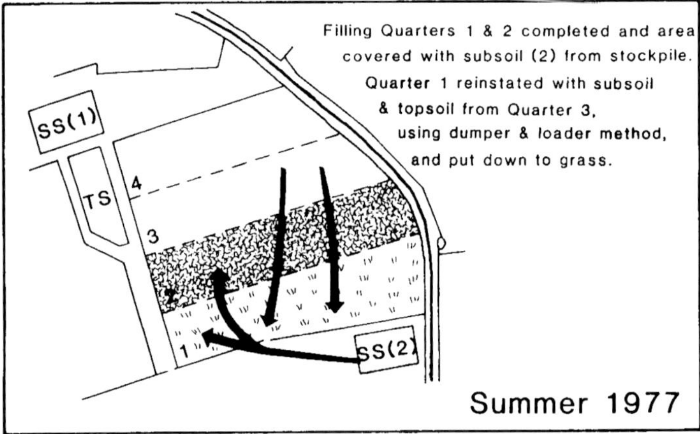

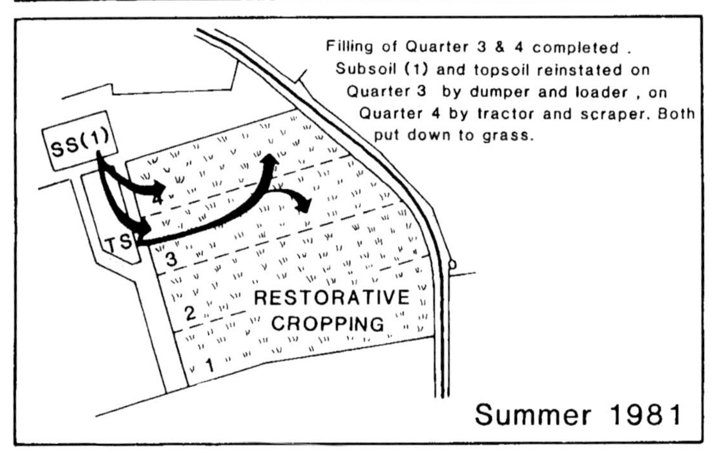

Fig. 69 Operations plan for the Bush Farm Agricultural Restoration Experiment. Initial, intermediate and late restoration stages. (Redrawn from Final Report (1982–1987) for Bush Farm, Upminster, Essex. Ready Mixed Concrete (United Kingdom)

Table 15 Methods of moving and replacing soil on the four quarters of the Bush Farm Experiment.

Quarter	*Area (ha)*	*Soil movement*	*Restoration method*	*Date of restoration*
1	1.68	Direct	Dumper	1977
2	1.96	Direct	Scraper	1978
3	1.94	Stockpiled	Dumper	1981
4	2.03	Stockpiled	Scraper	1981

compared with excavators and dumper trucks as in the previous example. The site of 8.2 ha was divided into four areas of roughly equal width. Each quarter was stripped in turn, quarried, infilled with industrial wastes and re-covered with subsoil and topsoil. A clay seal was included in the second half of the site. After installing underdrainage and taking an initial hay crop, a programme of restorative cropping was started. The complete process on the first quarter, from stripping the soil to the first hay crop, took three years; the fourth quarter was stripped in year 5 and yielded hay in year 7. Aftercare and monitoring of the whole site was then continued for a further five years. Figure 69 shows some of the stages in this stripping and restoration process.

Before work started, the agricultural land in this area was classified as Grade 2. At the end of the experiment in 1987, the best parts were graded 3a and the area most affected by gas emissions graded 3b/c. A small amount of settlement (0.4–0.6 m) occurred on the first quarter but was less in subsequent strips where the refuse was more compacted.

Table 15 shows the methods of soil handling used in each quarter. These provided valuable practical experience, but since each quarter had a unique combination of treatments, and was reinstated at different times, it was impossible to compare the crop yields statistically: one could not be sure that any observed difference in yield was due to any one particular factor, or, indeed, just due to chance. (In statistical parlance, the treatment effects were confounded.) The absence of an unquarried 'control' area from which crop yields were measured also limited precise conclusions. Nevertheless, yields of barley, wheat, peas and potatoes have compared favourably with local and national averages, and so the Experiment has been judged a success. It clearly is possible to restore landfill sites to productive use.

Not all land taken for gravel is prime agricultural land, of course, and it is questionable whether the Bush Farm type of restoration is always appropriate – especially now that farmers are being encouraged to take land out of agriculture through the Set-Aside scheme. A very different restoration scheme was adopted at Martin's Farm, St Osyth, by Essex County Council, again on an experimental basis. The aim here was to create a country park over a 24 ha landfill site with picnic areas, wildflower meadows with butterflies, and patches of woodland. In this instance, the County Council had taken over an already worked-out gravel pit and filled it in gradually with domestic waste over an 18 year period up till 1983. There was no stockpile of soil from the original excavation, so the final covering consisted of lorry loads of sandy clay subsoil, available free from various local sources, and spread over the compacted fill to a depth of about 0.6 metres. This is much less than the 1.2 metres of soil cover used at Bush Farm and elsewhere, but to have bought in soil would have been very expensive.

Although it might seem that the requirements for establishing native vegetation were much less demanding than those for arable crops, the scheme posed some interesting problems because of the poor soil properties and the constraints on remedial measures. It is true that, left to itself, such a site gradually becomes clothed with quite a variety of self-sown grasses and wildflowers; over 120 such species were recorded in the first four years after completing the soil cover. However, most of these were annual plants dependent on bare ground. Without periodic disturbance, creeping bent grass *Agrostis stolonifera* would soon dominate the area and render it very dull floristically. To create a meadow-like assemblage, it was necessary to cultivate the ground and sow it with a suitable selection of perennial grasses and wildflowers. It was then that the low soil fertility and unfavourable physical conditions made themselves apparent.

The poor soil structure and low organic matter content of the soil cover resulted in low water-holding capacity so plants suffered from drought in the summer. Dry conditions may have been exacerbated by heat produced through the decomposition of the refuse, as mentioned earlier, though methane did not appear to be a serious problem. Winter rainfall created the reverse situation because water could not drain through the compacted refuse, and there were no field drains as there were at Bush Farm. The soil therefore quickly became waterlogged, discouraging deeper rooting by plants, and thus limiting their ability to survive dry conditions. It was a classical 'chicken-and-egg' situation in which poor soil conditions inhibited plant growth while slow vegetation development limited the rate of soil improvement. This situation extended also to earthworms. Their rate of colonization depended on vegetation cover and was limited in the early years mainly to surface-living species such as the green worm *Allolobophora chlorotica* and *Lumbricus castaneus*. The deeper burrowing *Lumbricus terrestris* and *Aporrectodea longa* reproduce more slowly (chapter 5), and colonization was probably limited by the anaerobic conditions in the refuse.

Conventional treatments, such as ripping or mole draining, were not feasible because of the shallow soil cover, and, in any case, one of the aims of the experiment was to see if they could be avoided. Even the use of fertilizers at the outset risked the too luxuriant growth of creeping bent before the sown species were established. A rather low level of fertility is actually desirable if one wants to maintain floristic diversity (cf Wendlebury Meads, p. 137), and one is not too impatient about achieving results. In the event, 28 of the 31 sown species were successfully established within the first four years after sowing, producing a short, herb-rich sward in which several species of butterflies were able to build up strong colonies.

The establishment of shrubs and trees on this site is, at the time of writing, still undergoing trials to determine the best species and methods of planting. It will be an even slower process because of the limitations on deep rooting for many years. Eventually, trees will promote marked changes in the soil ecology by bringing up nutrients from the buried organic matter, and enriching the surface layers through the shedding of their leaves.

This chapter concludes with the description of an attempt to transfer and re-establish particular grassland communities using the same principles of soil handling as at Shepperton described earlier. The opportunity for this experiment arose during the construction of the M25 motorway near London Airport in 1980, when gravel was extracted from a small part of Staines Moor. Three small, discrete areas were located which supported plants characteristic of dry grassland, damp grassland and marsh: species such as bird's-foot trefoil *Lotus corniculatus* and lady's bedstraw *Galium verum* in the first case, jointed rush *Juncus articulatus* and creeping buttercup *Ranunculus repens* in the second, and yellow flag *Iris pseudacorus* and water forget-me-not *Myosotis scorpioides* in the third.

These areas were surveyed and pegged out on a 6 m grid, each square being bored with a soil auger to determine the nature and depth of soil and overburden. Some 400 m away a receptor pit, filled with clay and rubble five years previously, was graded to the appropriate levels, and in November 1980 the stripping and transfer took place. First the top 10–15 cm of vegetation and soil was scraped off and stored temporarily, then a second layer varying from 20 to 40 cm depending on the bore samples, and finally an underlying zone of unwashed sandy gravel, to give a total thickness of about one metre. These layers were straightway placed in correct sequence in the new area but in adjacent strips; at no time did vehicles travel on the undug donor site or on the placed material at the receptor site. The finished levels sloped down from the dry grassland zone to the marshland zone, and the correct water tables were maintained by pumping and by reference to a water level gauge.

Regrowth of the vegetation from roots and seeds was rapid, and by the end of May 1981 almost complete cover was re-established. After a couple more seasons with light grazing by cattle, the site, indeed, looked like a long-established meadow. Regular monitoring of the vegetation up to 1986 showed that all but one or two of the original 85 species had survived. This successful outcome demonstrated the unique nature of the topsoil as a living community, and its dependence, in this instance, on precise hydrological conditions. The secret of success lay in the ability to measure and re-create water tables and drainage appropriate to the different communities. The site received a reclamation award from the Sand and Gravel Association in 1988.

Whether or not such transposed communities can be considered natural or not is debatable. The purist will often accept the results of slow natural succession on ancient earthworks and mineral workings but not the deliberate manipulation of physical conditions to re-create communities lost through human impact. It seems that we take special delight in the surprise of Nature, rather than in the works of man. Certainly, such techniques are a two-edged sword. Can mineral extraction be opposed in areas of natural beauty if the landform and vegetation can be restored?

Each case must be judged on its merits in the light of existing ecological criteria and technological know-how. In any event, the restoration of a community is not a once-for-all process: suitable and sustained management is essential or else the habitat will change markedly. In general, we have much poorer knowledge of the soil fauna and how long this would take to re-establish itself. We can be pretty sure, though, that the rebuilding of ant-hills, for instance, would take decades, and some species that are dependent on ants might not be able to survive the intervening period. Clearly, we should not destroy rich habitats without very good cause even if we can re-create facsimiles. Nor, perhaps, should we sweep away or bury all the newer spoil heaps and quarries resulting from our industrial activities. These must increase the range of potential soil types, and could, in another 50 years, provide unique communities for us to study and enjoy.

Selected References and Further Reading

Agricultural Advisory Council on Soil Structure and Soil Fertility (Chairman N. Strutt; 1970). *Modern Farming and the Soil.* London: H.M.S.O.

Allen, H.P. (1981). *Direct Drilling and Reduced Cultivation.* Ipswich: Farming Press

Allinson, F.E. (1973). *Soil Organic Matter and its Role in Crop Production.* Amsterdam: Elsevier.

Anderson, J.M. (1978). Inter- and intra-habitat relationships between woodland Cryptostigmata species diversity and the diversity of soil and litter microhabitats. *Oecologia.* 32: 341–348

Anderson, J.M. (1988). Spatio-temporal effects of invertebrates on soil processes. *Biology & Fertility of Soils.* 6: 216–227.

Bal, L. (1970). Morphological investigations in two moder-humus profiles, and the role of the soil fauna in their genesis. *Geoderma.* 4: 5–36.

Ball, D.F. (1975). Processes of soil degradation: a pedological point of view. In: *The Effect of Man on the Landscape: the Highland Zone*; edited by J.G.Evans, S.Limbrey, & H.Cleere. Research Report 11; 20–27; Council for British Archeology. [Contains other soil/archeology papers of interest].

Ball, D.F. (1986). Site and soils. In: *Methods in Plant Ecology,* edited by P.D.Moore & S.B.Chapman, 2nd edn. London: Blackwell.

Ball, D.F., Mew, G. & Macphee, W.S.G. (1969). Soils of Snowdon. *Field Studies.* 3: 69–107.

Ball, D.F. & Stevens P.A. (1981). The role of 'ancient' woodland in conserving 'undisturbed' soils in Britain. *Biological Conservation.* 19: 163–176.

Barr, J. (1969). *Derelict Britain.* Harmondsworth: Penguin books.

Barratt, B.C. (1964). A classification of humus forms and microfabrics of temperate grasslands. *Journal of Soil Science.* 15: 342–356.

Bather, B.E. (1980). *Soil Classification for Soil Survey.* Oxford: Clarendon Press.

Bauer, T. (1981). Prey capture and structure of the visual space of an insect that hunts by sight on the litter layer (*Notiophilus biguttatus* F., Carabidae, Coleoptera). *Behavioural Ecology & Sociobiology.* 8: 91–97.

Bauer, T. (1982). Predation by a carabid beetle specialized for catching Collembola. *Pedobiologia.* 24: 169–179.

Bibby, J.S. (1986). Principal features of the formation of hill- land soils, their management and capability in cool, moist temperate climates. In: *Land and its Uses,* edited by F.T.Last, M.C.B.Holz, & B.G.Bell. New York: Plenum.

Black, C.A. (1986). *Soil-Plant Relationships.* London: Wiley.

Blower, J.G. (1985). *Millipedes: Keys and notes for the identification of the species.* Synopses of the British Fauna (New Series) No 35. London: Brill, for the Linnean Society of London & The Estuarine and Brackish Water Sciences Association

Bolt, G.H. & Bruggenwert M.G.M. eds (1978). *Soil Chemistry A: Basic Elements.* Developments in Soil Science 5A. Amsterdam: Elsevier.

Brian, M.V. (1977). *Ants.* London: Collins.

Burges, A. & Raw, F. eds (1958). *Soil Biology.* London: Academic Press. [Includes chapters on all major groups of soil organisms, including protozoa and Enchytraeidae not covered here].

Cameron, R.A.D., Jackson, N. & Eversham, B. (1983). A field key to the slugs of the British Isles. *Field Studies.* 5: 807–824.

Carson E.W. (1974). *The Plant Root and its Environment.* Charlottesville: University Press of Virginia.

Christien, E. (1978). The jump of the springtails. *Naturwissenschaften.* 65: 495–496

Coaker, T.H. (1965). Further experiments on the effect of beetle predators on the numbers of the cabbage root fly, *Erioschia brassicae* (Boucha), attacking brassica crops. *Annals of Applied Biology.* 56: 7–20.

Cooke, G.W. (1970). The carrying capacity of the land in the year 2000. In: *The Optimum Population for Britain,* edited by L.R.Taylor. Symposium of the Institute of Biology No 9. London: Academic Press.

Cooke, G.W. (1967). *The Control of Soil Fertility.* London: Crosby, Lockwood & Son.

Cooke, G.W. (1982). *Fertilizing for Maximum Yield.* 3rd edn. London: Crosby, Lockwood & Son.

Courtney, F.M. & Trudgill, S.T. (!984). *The Soil.* London: Edward Arnold.

Cruikshank J.G. (1972). *Soil Geography.*

Newton Abbot: David & Charles.

Curtis, L.F. Courtney, F.M. & Trudgill, S. (1976). *Soils in the British Isles*. London: Longmans.

Darwin, C. (1881). *The Formation of Vegetable Mould Through the Action of Worms*. London: John Murray. Reissued in 1945 as: *Darwin on Humus and the Earthworm*. London: Faber & Faber.

Davies, D.B., Eagle, D.J. & Finney, J.B. (1982). *Soil Management*. 4th edn. Ipswich: Farming Press.

Davis, B.N.K. & Coppeard, R.P. (1989). Soil conditions and grassland establishment for amenity and wildlife on a restored landfill site. In: *Biological Habitat Reconstruction* edited by G.P.Buckley. London: Belhaven Press.

Dempster, J.P. (1968). The control of *Pieris rapae* with DDT. I. Survival of the young stages of *Pieris* after spraying. *Journal of Applied Ecology*. 5: 451–462.

Department of the Environment, Ministry of Agriculture, Fisheries and Food, & Sand and Gravel Association. (1988). *Joint Agricultural Land Restoration Experiments: Final Report (1982– 1987) for Bush Farm, Upminster, Essex*. Feltham: Ready Mixed Concrete (United Kingdom) Ltd.

Dyke, G.V., George, B.J., Johnston, A.E., Poulton, P.R. & Todd, A.D. (1983). The Broadbalk Wheat Experiment 1968–78: Yields and plant nutrients in crops grown continuously and in rotation. *Rothamsted Report*. 1982(2): 5–44.

Edwards, C.A. & Heath, G.W. (1963). The role of soil animals in breakdown of leaf material. In: *Soil Organisms*, edited by J.Doeksen & J.van der Drift. Amsterdam: North Holland Publishing Co.

Eisenbeis, G. (1983). The water balance of *Trigoniophthalmus alternatus* (Silvestri, 1904) (Archeognatha, Malichidae). *Pedobiologia*. 25: 207–215.

Farb, P. (1955). *Living Earth*. London: Constable.

Fitzpatrick, E.A. (1980). *Soils, their Formation, Classification and Distribution*. London: Longmans.

Garner, H.V. & Dyke, G.V. (1969). The Broadbalk Wheat Experiment: The Broadbalk yields. *Rothamsted Experimental Station Report*. 1968(2): 26–49.

Gemmell, R.P. (1977). *Colonization of Industrial Wasteland*. (Studies in Biology No 80) London: Edward Arnold.

Gray, T.R.G. & Williams, S.T. (1971). *Soil Micro-organisms*. Edinburgh: Oliver & Boyd.

Greenwood, E.F. & Gemmell, R.P. (1978). Derelict industrial land as a habitat for rare plants in S. Lancs. (v.c. 59) and W. Lancs. (v.c. 60). *Watsonia*. 12: 33–40.

Haeck, J. (1969). Colonization of the mole (*Talpa europaea* L.) in the IJsselmeerpolders. *Netherlands Journal of Zoology*. 19: 145– 248.

Hall, D. (1982). *Medieval Fields*. Princes Risborough: Shire Archeology.

Hall, I.G. (1957). The ecology of disused pit heaps in England. *Journal of Ecology*. 45: 689–720.

Harding, P.T. & Sutton, S.L. eds (1985). *Woodlice in Britain and Ireland: Distribution and Habitat*. Huntingdon: Institute of Terrestrial Ecology.

Harley, J.L. & Smith, S.E. (1983). *Mycorrhizal Symbiosis*. London: Academic Press.

Hodgson, J.M. (1978). *Soil Sampling and Soil Description*. Oxford: Clarendon Press

Howard, Sir Albert (1940). *An Agricultural Testament*. London: Oxford University Press.

Hutson, B.R. (1980). Colonization of industrial reclamation sites by Acari, Collembola and other invertebrates. *Journal of Applied Ecology*. 17: 255–275.

Hyams, E. (1952). *Soil and Civilization*. London: John Murray

Jacks, G.V. (1954). *Soil*. Edinburgh: Nelson.

Jacks, G.V & Whyte, R.O. (1939). *The Rape of the Earth: A World Survey of Soil Erosion*. London: Faber & Faber.

Jackson, W. (1985). *New Roots for Agriculture*. Bison Publication. University of Nebraska Press.

Jenks, J. (1959). *The Stuff Man's Made Of; A Positive Approach to Health through Nutrition*. London: Faber & Faber.

Jollans, J.L. (1985). *Fertilizers in UK farming*. Reading: Centre for Agricultural Strategy Report No 9.

Jones, Dick. (1983). *The Country Life Guide to Spiders of Britain and Northern Europe*. Feltham: Country Life Books.

Jordan, C. & Smith, R.V. (1985). Factors affecting leaching of nutrients from an intensively managed grassland in County Antrim, Northern Ireland. *Journal of Environmental Management*. 20: 1–15.

Kerney, M.P. & Cameron, R.A.D. (1979). *A Field Guide to the Land Snails of Britain and North-West Europe*. London: Collins.

King, T.J. (1977). The plant ecology of ant-hills in calcareous grasslands. *Journal of Ecology*. 65: 235–256.

King, T.J. (1981). Ant-hill vegetation in acidic grasslands in the Gower Peninsular, South

Wales. *New Phytologist*. 88: 559–571.

King, T.J. (1981). Ant-hills and grassland history. *Journal of Biogeography*. 8: 329–334.

Kubiena W.L. (1953). *The Soils of Europe*. London: Murby.

Kühnelt, W. (1976). *Soil Biology: with special reference to the animal kingdom*. 2nd edn (translated by N. Walker). London: Faber & Faber.

Lavelle, P. (1988). Earthworm activities and the soil system. *Biology & Fertility of Soils*. 6: 237–251.

Lidgate, H.J. (1984). Benefits of fertiliser input to arable cropping. *Chemistry & Industry*. 18: 649–652.

Livingston, B.E. ed. (1922). *Palladin's Plant Physiology*. Philadelphia: P. Blackiston's.

Mackney, D. ed. (1974). *Soil Type and Land Capability*. Technical Monograph No 6. Harpenden: Soil Survey. (Contains papers on distribution of soil types, clay soils, and drainage).

MacRae, I.C. (1989). Microbial metabolism of pesticides and structurally related compounds. *Reviews of Environmental Contamination and Toxicology*. 109: 1–87.

Maltby, E. (1984). Response of soil microflora to moorland reclamation for improved agriculture. *Plant & Soil*. 76: 183–193.

Mellanby, K. (1967). *Pesticides and Pollution*. London: Collins.

Mellanby, K. (1971). *The Mole*. London: Collins.

Mellanby, K. (1981). *Farming and Wildlife*. London: Collins.

Miles, J. (1981). *The Effect of Birch on Moorlands*. Cambridge: Institute of Terrestrial Ecology.

Miles, J., Latter, P.M., Smith, I.R. & Heal, O.W. (1988). Ecological effects of killing *Bacillus anthracis* on Gruinard Island with formaldehyde. *Reclamation and Revegetation Research*. 6: 271–283.

Milner, C. & Ball, D.F. (1970). Factors affecting the distribution of the mole in Snowdonia. *Journal of Zoology*, London. 12: 61–69.

Moffat A.J. (1988). Forestry and soil erosion in Britain – a review. *Soil Use & Management*. 4: 41–44.

Morgan, R.P.C. (1985). Assessment of soil erosion risk in England and Wales. *Soil Use & Management*. 1: 127-131

Morrison, J. Jackson, M.V. & Sparrow, P.E. (1980). The response of perennial ryegrass to fertilizer nitrogen in relation to climate and soil. *Technical Report No 27*. Hurley: Grassland Research Institute.

Paul, E.A. (1984). Dynamics of organic matter in soil. *Plant & Soil*. 76: 275–285.

Phillips, J.M. & Hayman, D.S. (1970). Improved procedures for clearing roots and staining parasitic and vesicular-arbuscular mycorrhizal fungi for rapid assessment of infection. *Transactions of the British Mycological Society*. 55: 155–161.

Rheinheimer, G. (1985). *Aquatic Biology*. 3rd English edn. Chichester: John Wiley & Sons.

Roberts, R.D., Marrs, R.H., Skeffington, R.A. & Bradshaw, A.D. (1981). Ecosystem development on naturally-colonized china clay wastes. I. Vegetation changes and overall accumulation of organic matter and nutrients. *Journal of Applied Ecology*. 69: 153–161.

Rudeforth, C.C., Hartnup, R., Lea, J.W., Thompson T.R.E. & Wright, P.S. (1984). *Soils and their Use in Wales*. Bulletin No 11. Harpenden: Soil Survey. Harpenden.

Runham, N.W. & Hunter, P.J. (1970). *Terrestrial Slugs*. London: Hutchinson & Co.

Russell, Sir E.J. (1957). *The World of the Soil*. London: Collins.

Russell, Sir E.J. (1988). *Soil Conditions and Plant Growth*. 11th edn edited by A. Wild. London: Longmans.

Salter, P.J. & Williams, J.B. (1965). The influence of texture on the moisture characteristics of soils. II. Available-water capacity and moisture release characteristics. *Journal of Soil Science*. 16: 310–317.

Satchell, J.E. (1980). Soil and vegetation changes in experimental birch plots on a *Calluna* podzol. *Soil Biology Biochemistry*. 12: 303–310.

Satchell, J.E & Stone, D.A. (1977). Colonisation of pulverized fuel ash sites by earthworms. *Publicaciones centro Pirenaico Biologia experimental* 9: 59–74.

Scott Russell, R. (1977). *Plant Root Systems: their function and interaction with the soil*. London: McGraw-Hill.

Sheail, J. (1985). *Pesticides and Nature Conservation: the British Experience 1950–1975*. Oxford: Clarendon Press.

Shoard, Marion (1980). *The Theft of the Countryside*. London: Temple Smith.

Simpson, K. (1983). *Soil*. London: Longman.

Sims, R.W. & Gerard, B.M. (1985). *Earthworms: Keys and notes for the identification and study of the species*. Synopses of the British Fauna (New Series) No 31. London: Brill for The Linnean Society of London & The Brackish Water Sciences Association.

Smiles, D.E. (1988). Aspects of the physical environment of soil organisms. *Biology & Fertility of Soils*. 6: 204–215.

Soil Survey of Scotland (1984). *Soils and Land Capability for Agriculture; 1:250000 survey; Organisation and Methods*. Aberdeen: Handbook

8. Soil Survey of Scotland.

Sprent, J.I. (1979). *The Biology of Nitrogen-fixing Organisms*. London: McGraw-Hill.

Stamp, D. (1969). *Man and the Land*. London: Collins.

Stevens, P.A. (1975). Geology and soils. In: *Bedford Purlieus, its history, ecology and management*, edited by G.F.Peterken & R.C.Welch. Monks Wood Symposium 7. Huntingdon: Institute of Terrestrial Ecology.

Stewart, V.I. (1975). Soil Structure. *The Soil Association* 1: 6– 7.

Stonehouse, B. ed (1981). *Biological Husbandry: A Scientific Approach to Organic Farming*. London: Butterworths.

Strutt, N. (see Agricultural Advisory Council).

Tomlinson, T.E. (1971). Nutrient losses from agricultural land. *Outlook on Agriculture*. 6: 272–278.

Trafford, B.D. (1970). Field drainage. *Journal of the Royal Agricultural Society of England*. 131: 129–152.

Vine, A. & Bateman, D. (1981). *Organic Farming Systems in England and Wales*. Aberystwyth: University College of Wales.

White, R.E. (1987). *Introduction to the Principles and Practice of Soil Science*. 2nd edn. Oxford: Blackwell.

Wells, T.C.E., Sheail,J., Ball, D.F. & Ward, L.K. (1976). Ecological studies on the Porton Ranges: relationships between vegetation, soils, and land-use history. *Journal of Ecology*. 64: 589–626.

Worthington, T.R. & Heliwell, D.R. (1987). Transference of semi- natural grassland and marshland onto newly created landfill. *Biological Conservation*. 41: 301–311

Index